The Disordered Cosmos

A Journey into Dark Matter, Spacetime, and *Dreams Deferred*

Chanda Prescod-Weinstein

New York

Cover design by Pete Garceau
Cover image © iStock/Getty Images

Bold Type Books
116 East 16th Street, 8th Floor New York, NY 10003
www.boldtypebooks.org
@BoldTypeBooks

Printed in Canada

Originally published in hardcover and ebook by Bold Type Books in March 2021
First Trade Paperback Edition: May 2022

Published by Bold Type Books, an imprint of Perseus Books, LLC, a subsidiary of Hachette Book Group, Inc. Bold Type Books is a co-publishing venture of the Type Media Center and Perseus Books.

The Hachette Speakers Bureau provides a wide range of authors for speaking events. To findoutmore,gotowww.hachettespeakersbureau.comorcall(866)376-6591.

The publisher is not responsible for websites (or their content) that are not owned by the publisher.

Print book interior design by Trish Wilkinson

Library of Congress Cataloging-in-Publication Data

Names: Prescod-Weinstein, Chanda, author.

Title: Thedisorderedcosmos:ajourneyintodarkmatter,spacetime,anddreams deferred / Chanda Prescod-Weinstein.
Description: First edition. | New York : Bold Type Books, 2021. | Includes bibliographical references and index.
Identifiers:LCCN2020040651|ISBN9781541724709(hardcover) | ISBN 9781541724693 (ebook)
Subjects: LCSH: Astrophysics. | Particles (Nuclear physics). | Cosmology. | African Americans—Study and teaching. | Feminism. | Feminist theory. | LCGFT: Essays.
Classification:LCCQB461.5.P7352021 |DDC523.01—dc23
LC record available at https://lccn.loc.gov/2020040651

ISBNs: 9781541724709 (hardcover); 9781541724693 (e-book);
9781541724686 (trade paperback)

MRQ

2 2023

For Grandpa Norman z"l,
Grandma Elsa z"l, Grandpa Stanley z"l,
and Grandma Queens z"l.

For Uncle Cyril z"l.

For East LA and Brooklyn and Barbados and Hawaiʻi.
For Repiblik d Ayiti, where this freedom began.

we the black bodies understand each other at visible frequencies
without a dissection or death—which is to say witness
us the black bodies rejoice to become mortals again because
here is what is true:
a black body radiator be in thermodynamic equilibrium which is to say
a black body be at rest yes let the black bodies rest

—Lena Blackmon, from "quantum distributions for Sarah Baartman"

People need to know that we live in a universe that is bigger than the bad things that are happening to us.

—Margaret Prescod

CONTENTS

IN THE BEGINNING

A Bedtime Story

Once upon a time, there was a universe. We are not sure about how it started or whether there is a reason. We don't know, for example, if spacetime is ordered or disordered at the smallest scales, which are dominated by the weirdness of quantum mechanics. We are pretty sure that during the first sliver of a trillionth of a second it expanded very rapidly so that for the most part it looked the same in every direction and looked the same from every position. It was sameness everywhere. Except that particles started to blip out of nothing due to random fluctuations caused by quantum effects, maybe in spacetime, we are still not super sure about that. Then again, we are not super sure about this, either: for some reason those particles formed more matter than antimatter. That process, which formed a particle type called baryons, is called baryogenesis. From there those baryons started to form structures, and from those structures stars formed. Then the stars got old and some of them died in super epic, rather fabulous fashion. They exploded into supernovae, making heavy elements like carbon and oxygen in the process. Those elements went on to be the basis for all life on Earth. Earth is a planet, one of the structures that formed around stars from the leftovers of supernovae. Eventually, a smaller type of structure that we call life formed on Earth. Some of the life-forms that evolved were relatively hairless apes that use a variety of methods of communication. There are about 7 billion of these apes, with various levels of eumelanin and pheomelanin in their skin and hair, giving them a range of colors. The apes also have a lot of different hair textures. Some of the ones with less eumelanin have for a long time now been cruel to the ones with more, some of whom we know as "Black people." We know why this is although we don't fully understand the why, but it might be due to laziness or because they are jealous of our boogie. But despite this, Black lives come from the same baryogenesis, the same supernovae, and the same structure formation. No matter what the lowest-eumelanin people say, Black Lives Are Star Stuff and Black Lives Matter—all of them.

DESPITE THE FACTS OF THIS STORY, THERE'S STILL A LOT THAT we don't know about the universe. In science, we tend not to think in these terms—imagining the subject (us) and object (universe) to be distinct. This way of thinking is something we inherit from European thought, specifically the ideas of René Descartes. When we study the Andromeda Galaxy, we record its details as Cartesian thinkers, seeing it as something apart from ourselves and our home in the Milky Way. But at the same time, we are in a very technical sense bound up with Andromeda. It has its own story: it is the Milky Way's nearest major neighbor, and its existence does not trace back to a common origin with the Milky Way. Yet, in the future, these two galaxies will merge because they are bound together in a gravitational potential, which we can think of as a well in which they are both slowly spiraling downward, destined to eventually meet. Don't worry—this collision isn't set to be fully underway for another 4 billion years, and it won't be the kind of violent chaos that we imagine when we think about the word "collision." This isn't two cars smashing into each other, quickly and violently. Rather, it is stars and gas and (maybe?) dark matter particles reorganizing themselves into a new formation, guided by their gravitational relationships with one another.

This story is maybe our story. I say maybe because around the time that this collision occurs, our sun will be dying and our solar system will be destroyed in its death throes. Before its life ends completely, the sun will expand the amount of space it takes up, changing what constitutes the habitable zone of this solar system before eventually destroying the earth entirely. By then, we may have self-immolated anyway, but perhaps we will have just relocated to another solar system in a galaxy far, far away, using technology that is completely unimaginable and even unbelievable to me now. Or perhaps we will be in a solar system closer by, still in the Milky Way, in which case we will be carried along with the collision. The observations that our progeny will use to watch this phenomenon slowly unfold over the course of millions of years will require careful calculations about their location relative to all of the action.

As it is, we already do this. We are always studying our location in the universe, even when we tell ourselves that we are simply looking outward, beyond ourselves. In our attempts to learn more about the structure of galaxies, we spend an enormous amount of time looking at our own and wondering if it is normal. We are still unsure whether the Milky Way is an average spiral galaxy or whether there is something special about it. Even though we are not the center of the universe, because indeed the universe has no center, we are the reason that we bother with the universe at all. Our location in all of it matters.

Some of us wonder about where we belong more than others. I am a descendant of Indigenous Africans whose connection to the land was forcibly severed through kidnapping and the colonization of their bodies. West Africa is enormous and full of so many different peoples. I do not know and will likely never know for sure which Indigenous communities my ancestors came from, so the question of location remains fraught for me. I am also by and for East LA (east of downtown Los Angeles), and forged from Black American, Black Caribbean, Eastern European Jewish, and Jewish American histories. Today I split my time between where I live on the Seacoast region of New Hampshire and where my spouse lives in Cambridge, Massachusetts. Los Angeles, Cambridge, and New Hampshire are colonial names for the homelands of the Tongva, Pennacook, Wabanaki Confederacy, Pentucket, Abenaki, and Massachuseuk Nations. These locations and the people rooted in them matter in this universe too.

I am also a scientist who as a child terrorized her single mother by persistently questioning everything. I am a born empiricist, someone who by nature (ask my mother!) takes seriously that information should be collected and then provided as a mechanism for explaining why the world is organized in the ways that it is. This commitment to rationalizing order often seemed to center on my household chores, but I also wanted to know why mathematics so accurately described the universe and how deep that relationship goes. That question, along with the need to have some kind of career because I knew that bills had to be paid somehow, is why I decided at age 10 to become a theoretical physicist. It is also a question that remains the subtext of my work as a theoretical physicist nearly 30 years later.

But I also wanted to know why my third-grade teacher had left all the Black children with two Black parents off the playbill for our class's forthcoming modernist production of *Strega Nona* (produced and directed by actress Conchata Ferrell z"l), where I was to play one of Macbeth's three witches. Mrs. M threw me out of class for asking the question, but at the time I didn't think of it as a challenge to her authority. I simply wanted to know if she was a racist. I was curious. I had watched my mom's grassroots organizing combating racism and sexism, I had experienced racism by her side trying to get motel rooms on road trips, and I just wanted to know if I had discovered an instance of it on my own.

When I was 10, I thought that I could keep my curiosity about the mathematics of the universe and the existence and function of racism separate. But it was not to be. A hard lesson I learned as I emerged from my mother's home into rarified academic settings (first stop: Harvard College) was that learning about the mathematics of the universe could never be an escape from the earthly phenomena of racism and sexism (and now that humanity is moving deeper into our solar system, racism and sexism are no longer earthbound). As I progressed through college, graduate school, and teaching, I learned quickly and painfully that physics and math classrooms are not only scenes of cosmology—the study of the origins and inner workings of the physical universe—but also scenes of society, complete with all of the problems that follow society wherever it goes. There is no escape.

In physics, matter comes in different phases. For example, water and ice are different phases of the same chemical—liquid and solid. A phase transition occurs when matter changes from one phase to another. We see such a phase transition occurring when water evaporates: the liquid becomes gas. When it freezes, the transition is from liquid to solid. Phase transitions also occur in environments that feel far less mundane to us, for example, when a massive star goes supernova and converts from plasma to a neutron star that is some combination of superfluids and solids quite unlike those we find on Earth. Similarly, I had to undergo incredible intellectual phase transitions to conceive of what it meant to go from being a Black girl who loved but did not understand particle physics to a queer agender Black woman who loves—and is one of the chosen few to understand how much we don't understand—particle physics. My new understanding that society would follow me into the world of physics was also something of a phase transition for me.

This book will reflect these different phases in order to provide a holistic picture of the ways of knowing that we call particle physics and cosmology. I used to think physics was just physics, separate from people. I thought we could talk about particles without talking about people. I was wrong. At different points I came to understand physics as something that involved people, and that particular understanding has gone through different phases of its own. Studying the physical world requires confronting the social world. I know personally that social barriers impact the practice of science, its results, and the people who comprise the community we call "science." In this book, I will reflect to readers both my love for science and the difficulties people like me face in holding on to that love. For this reason, what follows is broken into four phases: *Just Physics*, *Physics and the Chosen Few*, *The Trouble With Physicists*, and *All Our Galactic Relations*.

This book is also part of a long tradition of scientists taking a moment to share with the wider world how they see science. Historically, scientists have aimed to give readers a sense of what communications researcher Alan G. Gross calls "the scientific sublime"—a feeling of awe at the universe and our place within it. This was a hallmark of Carl Sagan's science communication style, and I think it is why his documentary and book *Cosmos* captured the world's imagination and sustained me through difficult moments during college. Almost always, the scientists who have had the opportunity to share their views on science have been white men. Necessarily, as a Black agender woman, I see science differently than my science communication ancestors have because contrary to the usual lore, who you are matters in science. When you're looking at the world from the margins, a persistent feeling of "the sublime" can feel out of reach as you struggle against mundane and pervasive forces of oppression. It may, therefore, be tempting to cast this book as radically outside the popular science genre because I go beyond the sublime to acknowledge the big role that social phenomena play in science. Some will point to my own life as the central narrative of this text, but while you will learn some things about me along the way (and other scientists too), I am not the point. Much more interesting is the question of how we get free.

What does freedom look like? When I put this question to artist Shanequa Gay, she told me, "Freedom looks like choice making without having to consider so many others when I make those choices." I hear in Shanequa's response a deep cry for space to self-actualize, to not always be stuck in survival mode. A sketch of Shanequa's painting, *We Were Always Scientists*, appears at the beginning of this book; I commissioned that painting partly because I was trying to figure out my own answer to this question. I asked Shanequa to envision unnamed Black women scientists under slavery. I wanted to challenge the idea that "scientific thought" has been the exclusive purview of Euro-Americans and those of us who have been trained in their knowledge systems. I also wanted something to remind myself that I belonged in my physics department office, and to remind myself that even in the worst conditions, Black women have looked up at the night sky and wondered.

Those women whose names I do not know, who may or may not be part of my bloodline, are as much my intellectual ancestors as Isaac Newton is. In fact, it is through the lessons those women passed on that I have learned to manage living with the Isaac Newtons of the world: those who are good at physics, but who are not good to people. These ancestors also serve as a reminder that the universe is more than our attempts to manipulate it. I don't have to end up like Newton, who served as warden of The Royal Mint in the late 1600s and was said to enjoy his ability to burn at the stake, hang, and torture coin counterfeiters. I don't have to end up like J. Robert Oppenheimer, the brilliant and tragic theoretical physicist who oversaw the creation of the first nuclear weapons and spent the rest of his life trying to undo the damage. I believe we can keep what feels wondrous about the search for a mathematical description of the universe while disconnecting this work from its historical place in the hands of violently colonial nation-states. With this book, I hope to map out for myself and for others an understanding that creating room for Black children to freely love particle physics and cosmology means radically changing society and the role of physicists within it. In the end, I have two big dreams for Black children and others, besides clean water, good food, access to health care, and a world without mass incarceration:

1. To know and experience Blackness as beauty and power
2. To know and experience curiosity about the night sky, to know it belonged to their ancestors

That, too, is freedom.

בָּרוּךְ אַתָּה יי אֱלֹקֵינוּ מֶלֶךְ הָעוֹלָם אֲשֶׁר בִּדְבָרוֹ מַעֲרִיב עֲרָבִים. בְּחָכְמָה
פּוֹתֵחַ שְׁעָרִים, וּבִתְבוּנָה מְשַׁנֶּה עִתִּים וּמַחֲלִיף אֶת הַזְּמַנִּים, וּמְסַדֵּר אֶת
הַכּוֹכָבִים בְּמִשְׁמְרוֹתֵיהֶם בָּרָקִיעַ כִּרְצוֹנוֹ. בּוֹרֵא יוֹם וָלַיְלָה, גּוֹלֵל אוֹר מִפְּנֵי
חֹשֶׁךְ וְחֹשֶׁךְ מִפְּנֵי אוֹר. וּמַעֲבִיר יוֹם וּמֵבִיא לָיְלָה, וּמַבְדִּיל בֵּין יוֹם וּבֵין לָיְלָה.
יי צְבָאוֹת שְׁמוֹ: אֵל חַי וְקַיָּם תָּמִיד יִמְלוֹךְ עָלֵינוּ לְעוֹלָם וָעֶד. בָּרוּךְ אַתָּה יי,
הַמַּעֲרִיב עֲרָבִים

Blessed are You, Universe, who brings in the evening with a word, in wisdom opening the gates and with understanding changing the times and seasons, ordering the stars along their paths in the sky. Creator of day and night, rolling back light from the dark and dark from the light, You make the day slowly fade and bring in the night, dividing between day and night, how great is Your Name. Living Universe, may we always feel your Presence in our lives. Blessed are You, Adonai, who brings in the evening.

PHASE 1
JUST PHYSICS

In which the universe is, for a time, human-free.

ONE

I ♥ QUARKS

THE STORY GOES LIKE THIS. ME: BLACK CHILD ON A SCHOOL bus that is slowly crawling along the 10 Freeway East, windows down, exhaust filling her nose and lungs, causing headaches that stop only years later when her dreams of particle physics carry her far away from the Los Angeles smog. I am reading and then taking breaks from reading to tell whoever is left on the bus—just a handful of children because between my school's grades six through twelve, only about four of us live this far away—about these things called quarks. I don't know what a quark is or where the name comes from. I don't particularly care about the name either. But I know that the world is made out of quarks. I know that my brain is a quark and electron collection.

These particles are not just a Black child dreaming. The Standard Model of particle physics is also all the things that a Black child is made out of. It is all of the things I am made out of as a scientist who has reached adulthood, still fascinated with what she still doesn't understand. The journey to know quarks in a more technical sense than *A Brief History of Time* could provide had an almost dizzying number of twists and turns because the math that describes them is some of the most complex in all of particle physics. There were many pieces I had to understand first. And instead of being put off by my descent into the world of particles, my attraction to them deepened with each step.

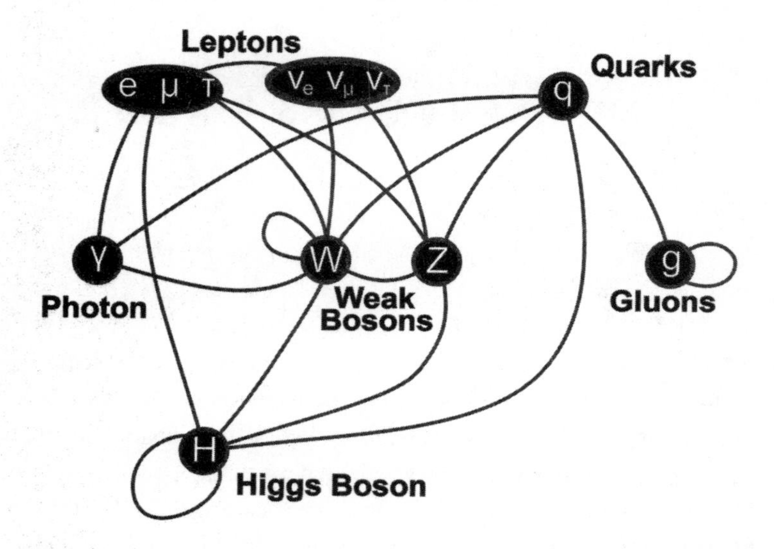

Figure 1. This diagram, made by a Wikipedia contributor, gives you a loose picture of the particles in the Standard Model and how they interact with each other. The top row of bubbles (leptons on the left and quarks on the right) are the matter particles. The middle row of bubbles (the photon, W and Z bosons, and gluons) are the force mediating particles. The bottom row is the Higgs boson. The lines between the bubbles show some of the possible interactions between these particles.
TriTertButoxy

During my intellectual adventure, I learned that I particularly enjoy a neatly ordered tale of an organized universe that can come off like a delicately constructed sum of its parts. This Standard Model is the framework that we use to describe and make predictions about elementary particles, the fundamental building blocks of matter, and three of the four known forces of the universe—electromagnetism, the weak force, and the strong force. Gravity, the one force we've been unable to fit into the Standard Model, has taught us that forces are felt—completely surrounding us—but never seen. This is in part why gravity doesn't fit in, because it is embedded in spacetime, which literally completely surrounds us. By contrast, the other fundamental forces are in fact mediated—communicated—by a special class of particles called vector bosons.

Before we get to what I mean by "vector bosons," let me add that they are only one of the many strange new things that come up in particle physics. Every realm of intellectual work has its own vocabulary, and I may be biased, but I guarantee you that few if any are as strange, fantastical, and apparently true as in the world of particle physics. For example, there's spontaneous symmetry breaking, a phenomenon where the equation that governs a particle's behavior obeys certain rules, but then when you actually use the equation to calculate the particle's least energetic behavior, the particle doesn't appear to obey those rules. What? Particle physics is full of mathematical stuff like this that seems completely unreal, and yet, all of our experimental data matches these strange mathematical ideas. Though I work at the intersection of astrophysics and particle physics, it is particle physics that continues to teach me over and over again that the universe is always more bizarre, more wonderfully queer than we think.

Particle physics, for me, isn't just about organizing information, although I admit that I get a certain pleasure from that. It's also about the basic premise of what all physicists do. Whether we are studying particles or new ways to make powder stick to people's faces (yes, that's a job you can have!), we study systems as they change in time and look for patterns or make predictions about patterns in their behavior. We used to think that absolute predictions could be made if we had sufficient information. One of the toughest lessons of the twentieth century was that this is not the case—our world is at base quantum physical in nature.

For our purposes here, quantum physics (which physicists call "quantum mechanics") means that the fundamental properties of particles are such that we must now understand that each event in the universe is but one probability among others. Some events are more likely, but everything is possible. The probabilistic nature of quantum mechanics is particularly noticeable in the land of particle physics, where the fundamental objects are very small and more evidently governed by quantum rules. We can never really predict exactly what particles will do, but we can calculate the likelihood that something will happen and the timescale over which we expect it to occur. The quantum world of particles requires a kind of stretching of our scientific imagination into our scientific reality: things that do not seem intuitive are now what is real. The existence of any given object in our everyday life seems definitive, guaranteed. The table my feet are resting on is there—except there's an incredibly small, almost zero probability that in a moment it won't be.

That's not all of the strangeness of particle physics. Here is some more: all particles fall into one of two quantum mechanical categories, one of which is the boson (named after Indian physicist Satyendra Nath Bose). The other is a fermion (named after Italian American physicist Enrico Fermi). The difference between bosons and fermions comes down to a rule called the Pauli exclusion principle (named for Austrian physicist Wolfgang Pauli) and a quantum mechanical property that each particle has, which we call spin. The value of quantum spins are measured in multiples of a number we call Planck's constant. Bosons have the property that their quantum spins always occur in whole numbers (zero, one, two, etc.). The famed photon, the particle that mediates electromagnetic interactions and that we experience as light is a vector boson, a particle with a spin of one. Fermions, which include my beloved quarks, have the property that their spins always occur in multiples of a half (1/2, 3/2, etc.). The electron has a spin of a half and so do all of the quarks.

Importantly, this spin property is not like a spinning ball, but it gets its name because it is part of the quantum mechanical counterpart to our everyday notion of rotation. This quantum spin is additive: any object that contains particles has a spin, because the individual spins of those particles will combine in interesting ways dictated by the rules of quantum mechanics. Therefore atoms, which are made out of electrons and quarks, also have a quantum spin. This quantum spin has significant consequences for the structure of matter. This is a result of the fact that bosons and fermions obey different rules, which tell us how the particles can be distributed into different quantum energy levels. I tend to think of bosons as pep squad particles: they are happy to all share the same quantum energy state together. Fermions? Not so much. They're grubby and don't share well. More technically, they have to obey the Pauli exclusion principle, which says that no two fermions can share the same quantum state in the same quantum system at any given time. Students who take high school chemistry often struggle with figuring out orbitals, where electrons can only go into a few slots before the slots are full and new electrons have to move to higher orbitals. This is Fermi statistics and the Pauli exclusion principle in action.

As complicated as this sounds, the idea of spin is often taught to sophomore and junior physics majors in college. We are partly able to do this because we don't spend any time trying to make sense of why particles come in these two distinct formats. It is instead a definition that students are encouraged to accept as a fundamental axiom, and if they'd like to think deeply about that, there's maybe a philosophy class they can take in another department. The bizarre probabilistic math of quantum mechanics makes predictions that match our experiments, and for many physicists, that's enough since developing models that match data and testing theories using data are the two primary activities of physicists. Many of us accept quantum physics without ever trying to make sense of it, and we're not particularly encouraged to either. In other words, you're doing no worse than a future professional physicist if you just accept what I'm saying here without any substantive understanding of why things are this way. Of course, the question of "why" constantly lurks in the background, but research on what is called "foundations of quantum mechanics" has largely been relegated to the margins of physics research, in part because it's difficult and in part because it hasn't been profitable. There's still a lot that we don't know.

We do know that everything we have ever seen in the universe is made out of two categories of fermions: quarks, which like to combine to form other particles, and leptons, which like Bartleby the Scrivener, would prefer not to. Everything we have ever seen in the universe that is made out of quarks and leptons has a mass because of the Higgs, a scalar boson (which has a spin of zero). Every force that we know of in the universe, except gravity, is mediated by vector bosons, like the gluons that are responsible for the way quarks glue together into other particles. Some physicists think gravity may be mediated through spacetime by a spin 2 boson called a graviton, but for now that is a hypothetical idea that goes beyond the Standard Model of particle physics.

The Standard Model itself is largely understood to be experimentally complete as of 2012. Before then, we had detected elementary particles in three categories: quarks, leptons, and vector bosons. Many of these detections happened in particle colliders—literally experimental setups where particles are shot at each other and then the equipment picks up the signatures of the pieces of the smashed-up particles. But we still didn't know for sure how these particles got their mass. In 2012, scientists announced that this technique had yielded definitive experimental evidence for the hypothesized Higgs boson. The discovery of the Higgs (named for British physicist Peter Higgs), which gives *most* other Standard Model particles their mass, meant that for the first time, humanity had direct evidence for a fourth category in the Standard Model: scalar bosons. With this, the major predictions of the Standard Model were fully affirmed by experimental data. As I write, the Higgs is the singular scalar boson in the category of observed particles.

And recall, the Standard Model is everything we've ever seen. Maybe it is everything we will ever see. There's a kind of power in being able to name . . . everything. Here is how physicists use it: my faves, the quarks, get their name from a line in Irish modernist writer James Joyce's *Finnegans Wake*, "Three quarks for Muster Mark!" Quarks come in six types: up, down, charm, strange, top, and bottom. The top and bottom are also sometimes called truth and beauty, respectively. According to historian Michael Riordan, these names were preferred by some physicists because they were looking for "truth" and "naked beauty." Ultimately the pairing of "top" and "bottom" prevailed. Quark is a departure from the standard particle nomenclature, many of which have an -on at the end, in reference to the electron (from the Greek word for amber, *ἤλεκτρον*, elektron), or diminutive -ino, used in Italian to mean "small."

The electron is part of the lepton family, which also includes the muon, the tau (sometimes tauon), and the known neutrino family*: an electron neutrino, a muon neutrino, and a tau neutrino. In the vector boson category, we have the photon, gluon, Z boson, and W boson. The lepton and quark families, along with the W boson, all have antimatter counterparts. The antielectron is better known as a positron, and every quark has an antiquark. These partner particles are completely similar except that they have the opposite charge sign, e.g., the electron is negatively charged and the positron is positively charged. These particle partners are also completely dissimilar in that leptons and quarks, not their antipartners, dominate matter. Why? We have no idea. Answering this question is a major area of research in particle physics.

As exciting as quarks and leptons are, you're more likely to hear about neutrons and protons. I'm completely aware that neutrons and protons (known collectively as nucleons because they appear in all atomic nuclei) have no feelings, and yet, I sometimes

---

* Neutrinos, for some reason, don't get their mass through the Higgs mechanism in the same way that other particles do. More on them in the next chapter!

feel a little sad for them. They are not elementary particles at all, but rather composites made out of quarks. The neutron is a particle with no charge, and it is made out of an up quark and two down quarks. Protons by contrast are positively charged and are made out of two up quarks and one down quark. The difference in charge comes down to the difference in quark composition. If we use the charge of the electron as our standard for comparison, up quarks have a charge of two-thirds (of an electron charge), and down quarks have a charge of minus one-third. So a little arithmetic (2/3 + 2/3 - 1/3 = 1) shows that a proton, made of two ups and a down, will give us a total charge of +1. The neutron is literally neutral because its quark charge arithmetic is 2/3 - 1/3 - 1/3 = 0. While it's true that neutrons and protons sometimes feel like they are faking it—masquerading as fundamental particles while actually being composite particles—they do reflect the properties of their underlying building blocks. This means that just as there are antiquarks, there are antiprotons. There is even antihydrogen, which has been observed in the lab!

It's worth remembering that we didn't always know that protons and neutrons were made out of quarks. Before we knew about the possibility of quarks, one of the open questions in physics was why the proton and neutron had the charges they did. The quark family, particles first conceptualized by Murray Gell-Mann and George Zweig and then expanded by James Bjorken, Sheldon Glashow, James Iliopoulos, Makoto Kobayashi, Luciano Maiani, and Toshihide Maskawa, were eventually observed as a phenomenon in experimental physics, giving an answer to this question of why. Nucleons and other baryons—the class of particles made of three quarks—have the charge that they do because they are made of quarks that have specific charges.

Anyone who has dealt with a curious toddler or an obstinate teenager knows that the next question is still: why? One definition of a physicist is that a physicist is a person who gets really bugged about a question like this and then spends their life trying to solve it and problems that are related to it. Why do quarks have the charge that they do? We don't know—that's an open question. Want to know why something happens? That's cool, but the universe doesn't really care about our feelings. What's great about physics, and what made it intriguing to me as a child, is that physicists have come up with a really interesting way of gathering information about our universe that leads to a neatly self-contained story about how all of the pieces fit together—even though the universe isn't necessarily easy to understand. What we can say about quarks, therefore, is not *why* their properties are the way they are, but we have developed a complex and beautiful mathematical theory that accurately accounts for all of the properties of quarks.

How? Nuclear physicists had detected the existence of eight kinds of mesons—unstable, composite particles that we now know to be made of one quark and an antiquark—and were trying to understand the underlying structure that produced this phenomenon. An explanation was simultaneously realized by Yuval Ne'eman, who was born in Tel Aviv, Mandatory Palestine, and was a professor at Tel Aviv University in Israel; George Zweig, a Soviet-born, California Institute of Technology (Caltech) PhD who was working at the European Organization for Nuclear Research (CERN); and Murray Gell-Mann, a Manhattan-born physicist who was a professor at Caltech. What this international group realized was that mesons could be explained by organizing them into a model that had new particles in it—quarks. The recognition of quarks as a component of the model was motivated by the exploration of a sophisticated mathematical symmetry.

In physics, a system has a symmetry related to one of its properties if that property doesn't change even if the system is somehow altered. A very simple example of this is rotating a spherical ball: it looks the same even if you alter the ball's circumstances by turning it in your hands. This is rotational symmetry. Symmetries are valuable because they help us make approximations so that it's easier to solve the math problems that arise in physics.

Physicists are taught from day one of frosh physics that if you can simplify a problem to something that has some symmetries, you should. In fact we have a running joke that all of physics boils down to "approximate this cow as a sphere." Cows don't have a shape that is easy to work with from the perspective of solving physics equations, but spheres? We know how to calculate with spheres because they are relatively simple geometric objects that display rotational symmetry—they look the same no matter how we turn them around. This symmetry effectively removes a level of complication, a dimension from the equations we might otherwise have to solve, whereas cows only have a left-right symmetry that doesn't simplify things much for us, and even that is an approximation because their left and right sides are probably somewhat different. So, we reduce the cow to a sphere for the purposes of doing physics.

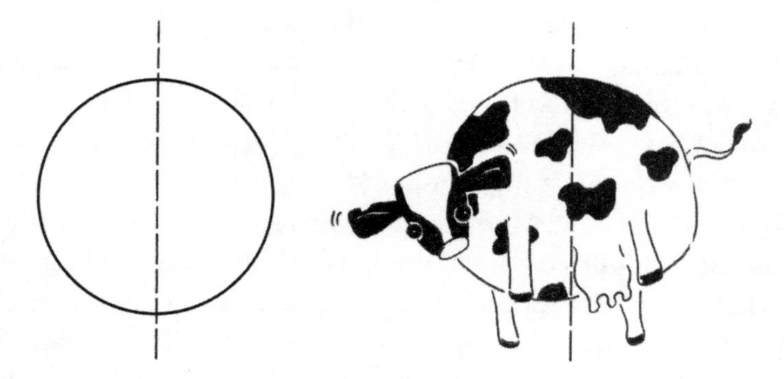

Figure 2. In this figure, you can see that the cow—though it is quite round—looks different if you look at the left half and the right half. By contrast, the sphere looks exactly the same no matter which side you choose.
S. Zainab Williams

In the twentieth century, symmetry was a deeply important guiding principle for particle physics, in part thanks to the work of German mathematician Emmy Noether. Her theorem teaches us that when a system has a symmetry, there is an associated conserved quantity, a property of the system that doesn't change with time even as the system evolves. This rule provided a road map for physicists: where there is a symmetry in the math, look for a conserved quantity. It also suggested that symmetries were mathematically foundational to descriptions in particle physics. Gell-Mann and others were following this rule when they developed what has come to be known as quantum chromodynamics, or QCD for short. Quantum chromodynamics is a colorful name for a theory that uses color as an analogy for physical properties that have nothing to do with color. Unfortunately, QCD came of age in a time when Black people were still using a moniker given to us by white people—"colored"—and to this day, textbooks still sometimes call it "colored physics." A Black feminist physicist working in the 1960s would never have used this language, which—though it lacks the same social meaning—recalls for me the Third Reich–era phrase "Aryan physics." But Black people were almost completely excluded from particle physics until after the Civil Rights era, and a Black woman wouldn't even earn a PhD in any area of physics until the following decade (Willie Hobbs Moore, University of Michigan, 1972), nearly a century after the first Black man earned one (Edward Bouchet, Yale University, 1876).* What I'm saying is, I love the idea of QCD, but the language is a hot mess.

Quantum chromodynamics, like other parts of the Standard Model that are for no good reason whatsoever less interesting to me, is an extraordinary artifice—my favorite in all of particle physics. I love QCD both because it is mathematically rich, like piecing together an elaborate three-dimensional puzzle, and because solving a little problem with it leads to the creation of the dark matter candidate that is at the center of my research. But before I get ahead of myself (you'll have to wait until Chapter 2), there are

* Bouchet was also the first African American to earn a PhD from a university in the United States.

non-dark-matter-related reasons to like QCD. For example, each quark and antiquark partner has not just an electrical charge, but also something called a color charge. It turns out that there are three colors: red, blue, and green. Yes, the antiparticles have anticolor properties—they are antired, antigreen, and antiblue. These labels have nothing to do with the colors we are used to seeing with our eyes and everything to do with physicists enjoying naming things for any number of random reasons. For example, the axion, which we will discuss momentarily, was named after a laundry detergent, which took its name from the Greek Orthodox Church liturgy, which has a prayer to the Mother Mary called *Ἄξιον ἐστίν* or "axion estin." While the labels used to describe color charge are random, the phenomenon itself has a very specific importance: holding quarks together inside of mesons and baryons.

This property creates new rules that particles have to follow while they are exchanging the gluons that allow quarks to stick together in very particular configurations. It's almost like a game where the goal is to get a colorless particle at the end (and this is where the analogy suggests that not having color is apparently inherently good). There are two ways to achieve this colorlessness. In the case of the three quark baryons (like neutrons and protons), one quark of every color should be present to make the color charge white, or neutral. If you want to make a meson, you need two quarks—but not just any two quarks. One has to be a quark and the other an antiquark, and whatever color the quark is, the antiquark must be the anticolor of that color so that the two colors cancel out, leaving a neutral white charge.

For those reading between the lines for interesting analogies, there's actually nothing titillating to see in the language here. "Color" and "white as neutral" are here not as reflections of how the universe works, but rather how a homogeneous, white scientific community comes up with new names for stuff. Part of science, therefore, involves writing a dominant group's social politics into the building blocks of a universe that exists far beyond and with little reference to our small planet and the apes that are responsible for melting its polar ice caps. The color charge scheme makes about as much sense as naming the three charges chocolate, strawberry, and vanilla. A baryon could represent Neapolitan ice cream, and a meson would be ice cream–neutral. This maybe makes sense intuitively when you realize that mesons are extremely unstable, with lifetimes that are much shorter than a second. There are limits to this analogy at a technical level, I suppose. But what's important here is that color physics seems intuitive to physicists not because it's a great analogy for general audiences but because our educations socialize us into the "color + color + color = white" paradigm. It would be great if I never had to read the phrase "colored degrees of freedom" in a new scientific paper again.

But while I don't love the language, I have the privilege of understanding the math that undergirds it, and I know what a wonderful and curious thing quantum chromodynamics is. Particle physics often feels like the gift that keeps on giving because just when you think you've come to understand it, there's another layer. As I said earlier, the universe always turns out to be more complicated and queer than we think it is. For example, beyond color charge and spin, there are other quantum numbers: baryon number and isospin, among others. All of these numbers come together to lay out permutations of particles with a mix of properties, all of which unite to make the universe as we understand it.

There is something sublime about this ability to break complex things down into fundamental parts—this scientific reductionism. Indeed, particle physicist, quantum theorist, and feminist theorist Karen Barad points out in their book *Meeting the Universe Halfway: Quantum Physics and the Entanglement of Matter and Meaning* that "particle physics . . . is the ultimate manifestation of the tendency toward scientific reductionism." But Barad highlights this quality of our research in order to challenge this practice. They point out that quantum mechanics challenges the idea that the universe can be broken down into separately knowable parts, and that physicists have yet to allow this lesson to impact their approach to thinking through what we call "the most fundamental, basic objects in the universe." Barad also notes that, "quantum theory in all its applications continues to be the purview of a small group of primarily Western-trained males." While Barad doesn't argue that quantum theory, which they explain is the basis for the Standard Model, is dependent on the geography or gender of its practitioners, they do point out that how to interpret its meaning and implication for physics remains an open, unresolved question. Physicists may struggle to make progress in physics if we continue to draw within the lines of a reductionist framework.

I had a really hard time with Barad's comment the first time I read it specifically because they are complicating and even challenging what was so attractive to me about particle physics. It left me wondering how I might perceive all of this if I were not a product of an educational system rooted in Euro-American precepts. Would I still be fascinated by the layered hierarchy of particle physics, and all of its attending nomenclature? Or did my education give me the capacity to find particle physics deeply exciting, despite its juxtaposition with the nuclear weapons development that helped turn modern particle physics into a line of work? After all, I've known since I was a kid that particle physics was tied to the legacy of nuclear weapons. In fact, particle physics as a subdiscipline began as nuclear physics and only over time did the high-energy part—particles—split off from the low-energy part, which continues to be known as nuclear physics. Nuclear physics is now a crucial area of research for medical technologies, but it remains tied up in weapons and energy production whose safety standards, as outlined by Lacy M. Johnson in her book *The Reckonings*, are still fraught.

Worrying about this reminds me of the opening sequence of Cameron Crowe's English-language adaptation of Alejandro Amenábar's film *Abre los Ojos*, *Vanilla Sky*. In the scene, Tom Cruise's lead character David Shelby is driving down a familiar

New York City street against the backdrop of Radiohead's opening track to *Kid A*, "Everything in Its Right Place." The song, which Radiohead's lead singer Thom Yorke has explained is about depression, is an ominous foreshadowing of the dystopian direction that the film goes in. My experience with particle physics has something in common with this juxtaposition. I feel comfortable, extremely comfortable in fact, with how the Standard Model tends to locate particles in their correct mathematical structure, in their right place. But I am also low-key worried, on the regular, that finding comfort in this makes it difficult for me to see the larger physical picture, or perhaps is a refusal to see the larger picture. When I think about this, I too have Radiohead playing in the background.

In my heart, I fight with the history of the Standard Model of particle physics and the motivation behind it, but also every time I think I can't deal with physics or physicists anymore, it is the Standard Model that makes me stop in my tracks and think, "Wow." I get lost, in the best possible way, in the math—every single time. I will never get bored of picking up a particle physics textbook and starting from page one. At the same time, the history is replete with white men, few white women, few Asian people, few Black people, few Latinx* people (and among those, almost uniformly men), and no Indigenous North American people. No past nonbinary folk are known to the historical record. To this day, the

* To describe people who have origins in Central and South America, in this book, I have chosen to use the term "Latinx" rather than the conventional "Latino" in recognition of calls from nonbinary people and others for language that is non-gendered. I recognize that there is also an emergent term, "Latine," which some argue is a better replacement. My use of Latinx is not intended to actively stake a position on word choice as an outsider to this community discourse, but rather a decision to use a word that is more widely known at this point in time. This is a comprehension decision, rather than a political intervention. Importantly, Latinx people hold a variety of racial identities, including white. Here I use the term expansively, with the understanding that white Latinx people do not experience marginalization to the same extent that Latinx of color do.

community is still essentially like that. For example, the University of Toronto string theorist A.W. Peet is the only significantly accomplished and well-known nonbinary high-energy physicist (a broad term that includes particle physicists and quantum gravity theorists). I have no love for how my professional community is structured. But it's also the case that when I think about quarks, I experience the kind of loving hope that is best set to Def Jef's "Black to the Future": "We know where we're goin, because we know where we came from." Maybe, then, I'm not just a hack for colonial science, but more like Princess Shuri from *Black Panther*, giving particle physics a new spin and rhythm. This is not to say that the laws of the universe are not universal—but it may be that what we think we know is incomplete and will not be complete until we are able to think beyond how white men are trained to think in a Western educational setting.

Only time, and a community that does not have extensive barriers to the participation of people from a broad cross section of humanity, will be able to tell how our understanding of physics will change when our understanding of who can be a physicist changes. I am one of the first to confront and, to put it in quantum terms, tunnel through the barrier made out of the belief that it doesn't matter if Black women (people) are excluded or if parts of the universe are described by "colored physics," forever marking this work as "physics developed by and for white thinkers." It is unclear whether I am making it through because I have been assimilated or through the brute force of my own will and imagination.

Barad writes in their essay "No Small Matter: Mushroom Clouds, Ecologies of Nothingness, and Strange Topologies of SpaceTimeMattering" about the ways that quantum field theory—the calculational tool that undergirds all of particle physics—and the atom bomb "inhabit and help constitute each other." If I see my intellectual life as inhabited and constituted by QFT, as we call it, does that mean I, too, am inhabited and constituted by the atom bomb? I know the answer to this question. My discipline received extensive funding because of the perception that, like the Manhattan Project, particle physics could continue to serve state interests, both technologically and psychologically. As this possibility has diminished, so has our funding. The way I have inhaled particle physics enmeshes me with this historical trajectory. But I am still also one natural conclusion of a Black child dreaming of quarks—not because quarks could serve state interests, but because quarks nourished the soul. The Standard Model? It is how I fell in love for the first time.

TWO

DARK MATTER ISN'T DARK

SOME PARTICLES ARE BUILDING BLOCKS, AND THERE ARE others, like the neutrino, that are mostly for the end days. No, no I don't actually mean the Christian apocalypse. I mean that they are literally a common product of decay in the universe. In fact, Wolfgang Pauli first hypothesized neutrinos in 1931 because the numbers for certain radioactive decays weren't adding up, and a good way to explain the missing energy was to propose that an as yet undetected particle was carrying it away. Soon after Pauli proposed this idea, Enrico Fermi developed a theory of radioactive decay that took these new particles into account, and he named them the "neutrino," which in Italian means "little neutral one." Nearly thirty years after neutrinos were initially proposed, they were first observed by Clyde L. Cowan and Frederick Reines in what is known as the Cowan-Reines neutrino experiment at Savannah River Nuclear Reactor in South Carolina. These experiments tested the theory that an antineutrino produced in a nuclear reactor, when interacting with a proton, would react and produce a neutron and positron. The positron, the antimatter of an electron, would then be destroyed through contact with an electron, giving off two high-energy particles of light, gamma rays. The

Savannah River site experiments detected these gamma rays and the leftover neutron. The unique combination of two gamma rays and a neutron made clear that an antineutrino had been given off by the reactor, setting off the whole sequence of events.

In addition to being difficult to detect, neutrinos are kind of fabulous. They have no charge, but each type of neutrino has a charged leptonic partner. This means that they come in three flavors: the electron neutrino, the muon neutrino, and the tau neutrino. It took nearly 50 years to figure out that neutrinos have a mass. I was a senior in high school when that revelation first became public. Because their mass is so tiny, they are perpetually what we call "relativistic particles." They are able to move at speeds close to the universal speed limit, the speed of light. This makes them efficient carriers of energy away from, say, a site of nuclear decay. It's this feature that makes neutrinos extremely interesting from the point of view of not just particle physics but also astrophysics. One place neutrinos are made is in stars, and lots of them are made when stars explode—a phenomenon we call a supernova. Because of this, we use neutrinos, along with photons—particles of light—and the ripples in spacetime that we call gravitational waves to look at the universe. We are still unsure of what the neutrinos' masses are or how to explain why they have a mass that is extremely tiny but still bigger than zero. Everything we know about physics encourages us to expect the mass to either be zero or to be something notably bigger. And for a while, because we were unsure about their masses and whether they had any mass, we thought neutrinos were something called dark matter. It is only in the last decade or so that we have become certain that they aren't massive enough, leaving us with a problem. What the heck is dark matter?

Let me start here: dark matter isn't even necessarily real. "Dark matter" as a term was coined in French in 1906 by Henri Poincaré, who called it *matière obscure*. Twenty-two years before, English astronomer Lord Kelvin first proposed in 1884 that "many of our stars, perhaps a great majority of them, may be dark bodies." In the 1920s, Dutch astronomers Jacobus Kapteyn and Jan Oort similarly proposed the presence of something like the matière obscure based on observations of stars in the Milky Way and its galactic neighbors. In 1933, Swiss astrophysicist Fritz Zwicky proposed that there was evidence for what he called (in German) *dunkle Materie*, this time based on observations of clusters of galaxies. More evidence came from American* astronomer Horace Babcock in 1939, and by then the name "dark matter" had stuck even if it didn't really make sense because the problem wasn't that it was dark, but rather that it was unseeable, invisible.

This distinction is meaningful when we consider the first truly significant evidence for the existence of matière obscure, which came in the 1960s and 1970s, largely thanks to Vera Rubin's creative use of a new spectrograph developed by Kent Ford. This spectrograph breaks down light into different colors (more on this in Chapters 4 and 5), and Dr. Rubin was the first scientist to realize it could be used to measure galactic star speeds with unprecedented accuracy. The results showed that there was a significant mismatch between how fast stars were rotating around the center of the galaxy (if stars were the only matter in a galaxy) and how fast the stars were actually going. If all of the mass in a galaxy is contained in stars and dust, then we can measure how massive those galaxies are by looking at how much radiation we collect from the stars and dust. There's a nice physics equation that gives us a correlation between the luminosity—the brightness—and the mass. There's also another equation that gives us the relationship between the mass of the galaxy and how fast the stars are orbiting the galaxy's center. This is one of Newton's laws, and we teach it to

* "America" and "American" are words that can refer to people from anywhere in North America, Central America, and South America. Because this book focuses on the United States as its primary context, as a shorthand, I will typically use the word "America" to refer to the United States and "American" to refer to people who are from the United States.

high school students. But with galaxies, this is also where we run into a problem. The measured mass of all the stars together, based on the orbit speeds, does not match the measured mass of all the stars together based on the brightness measurements. The orbit speeds suggest there should be a lot more mass there.

This suggests, in turn, that there is a lot of missing, invisible matter. There are other possible solutions, like maybe our theory of gravity is wrong (and I'll touch on that later), but for now I will focus on the more popular idea that we need a solution to the missing matter problem because otherwise our two carefully collected types of data are not consistent with each other. Looking at star movements in galaxies was the first way that scientists understood that the missing matter problem was a real, big problem. But it is not the only way, and there are now several inconsistencies that can't be explained without the insertion of "dark matter" as part of the equation. In Chapter 6, I will discuss gravitational lensing, yet another example of evidence that there is an invisible matter wandering around the universe.

Existentially, dark matter is a reminder of how much we don't know about the universe. The Standard Model of particle physics can't make sense of everything. Thanks to a variety of astronomical measurements, we—that is to say, most cosmologists and particle physicists—think 80 percent of the matter in the universe is what has come to be known as dark matter. Our current understanding of the universe suggests that the constituents of everything we have ever seen—the very stuff that we are made of—only makes up about 20 percent of the matter in the universe. The rest is dark matter. And if, like Einstein taught us to, we expand our definition of matter to include energy, the composition breakdown becomes even more bleak: 5 percent Standard Model matter, 25 percent dark matter (whatever that is), and 70 percent dark energy (more on that later!). It turns out that the Standard Model is not all that, in the end. It may in fact only explain 5 percent of the matter-energy content in the universe. In other words, baryons, the Standard Model, the everyday stuff? Us? We are weird, completely abnormal. I don't just mean physicists. I mean all of us, including Sequoioideae (the redwood tree family), our entire planet, and our whole solar system. Space is mostly empty, and the parts that are not empty seem to be mostly filled with a kind of matter that is invisible to us. We haven't figured out whether there is a way for our scientific instruments to touch it; we don't know if it's one kind of particle or eleven hundred different kinds. Tim Tait's Venn diagram in figure 3 below shows some of the proposed particles, and again, we don't know if any of these provide the answer. What we do know is that it is responsible for holding our galaxies together and plays a key role in the formation of the stuff that we *can* see.

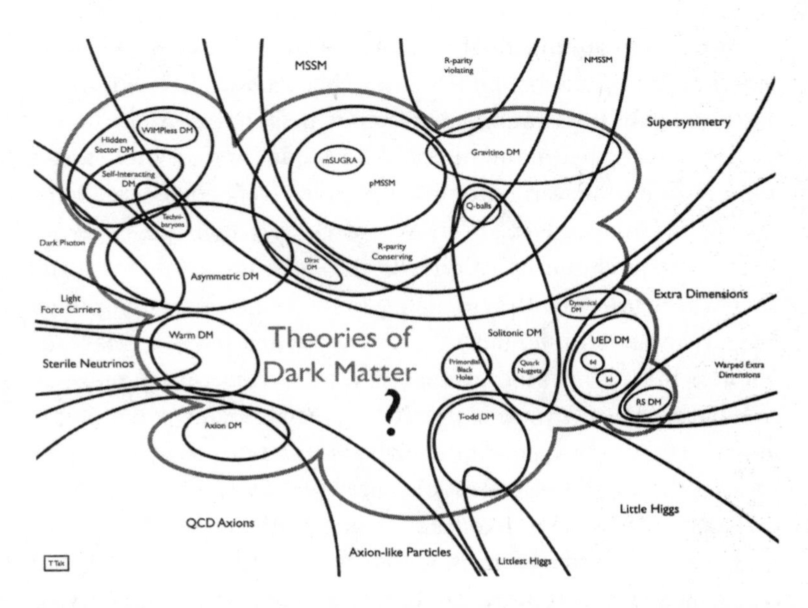

Figure 3. The greatest Venn diagram of all time: a representation of different theories of dark matter and how models overlap, for example, through shared properties. Particle physics theorist and professor at UC Irvine Tim Tait created this visual to help other physicists understand the array of possible dark matter models.
TIM M.P. TAIT

The most obvious question to ask here is why the dark matter particle doesn't just appear in the Standard Model. The answer: we don't get to choose the structure of the Standard Model. We are bound by the limits of the mathematical structures of our theories and by experimental data. The problem with dark matter is that we've never seen it, and there's no room in the Standard Model, which is built around what we have seen, for something we've never seen. Dark matter also has a public relations problem because it's got a bad name, literally. It should really be called "invisible matter," "transparent matter," or "clear matter." I vote for invisible or transparent matter because it doesn't remind me of a particularly bad era in Pepsi product management (for anyone younger than an early millennial, it's enough to say that "Crystal Pepsi," also marketed as "Pepsi Clear," had a massive marketing campaign—including a hyped Super Bowl ad—that ended in disaster for both Pepsi and one of Van Halen's beloved singles).

Of course, the first question was whether this could be explained using particles in the Standard Model, and for quite a long time, in fact until fairly recently, neutrinos were considered to be strong candidates. Neutrinos are not *entirely* invisible—they do have some kind of interaction with the electromagnetic force and therefore light—but the interaction is very small, making them largely invisible to us. But what we've learned in the last decade about neutrinos proves to us that they definitively cannot make up the majority of the invisible matter we need because they simply aren't massive enough. The neutrino would need to be hundreds or even thousands of times more massive than it is in order for it to be a good explanation for the missing matter problem.

Today, research into dark matter is considered to be "beyond the Standard Model" physics. The expectation is that this invisible matter is a particle we've never observed before. This kind of problem is again a defining one for a physicist: you can get so into it that you commit your life to it. That's not how I ended up becoming an expert on dark matter though. I fell into it because of the research opportunities I found available to me at a moment when I needed to be doing some kind of research and was a little bit lost. The orbit I ended up being captured in first was not what everyone was thinking about at the time, but rather something a little different: the axion.

The axion is still a hypothetical particle, but its conception as an idea is connected to yet another problem with the Standard Model. It turns out the Standard Model is not just missing a good candidate to explain the movements of stars in their galactic orbits; it has other problems too. One is called the strong CP problem, which is that the theoretical formulation of quantum chromodynamics does not break charge-parity (CP) symmetry. CP symmetry is the rule that the laws of physics are the same when a particle and its antiparticle are interchanged (this is charge symmetry) when it is viewed through a looking glass, as a mirror image (this is its parity symmetry). CP symmetry means that if you take a particle and replace it with its antiparticle and swap its left-right orientation, the physics works out the same. This symmetry is actually kind of fun to think about—consider swapping your hands for the antimatter version of your hands and then switching your left hand for your right hand. Under normal circumstances, this would pretty obviously change your life, but at the particle level, this symmetry is sometimes preserved.* The problem with preserving it in quantum chromodynamics, however, is the neutron electric dipole moment. The preserved CP symmetry does things to the neutron that shift it into a particle with new properties, properties that, as far as we can tell, are completely absent. The equations must

* Importantly, parity symmetry is not always conserved on its own, either. In 1956, Chinese American nuclear physicist Chien-Shiung Wu established that parity is not conserved in particle interactions governed by the weak force. Two theorists, Tsung-Dao Lee and Chen-Ning Yang, won the Nobel Prize for this idea because their work predicted this outcome. Wu did not have her essential experimental work recognized with a prize until two decades later when she was awarded the first Wolf Prize.

be shifted so that these tentacles are kept in check by breaking charge-parity symmetry.

Effectively, the magnificent Standard Model over-predicts things. The best patch for addressing this problem in the Standard Model is known as the Peccei-Quinn mechanism because it was developed by Roberto Peccei and Helen Quinn in the 1970s. Peccei and Quinn took a constant parameter in quantum chromodynamics and made it dynamical. Frank Wilczek and Steven Weinberg realized, independently, that this model had a new particle in it. This particle was called the Higglet by some scientists for a while, but now it is known by Wilczek's name for it, the axion.

Peccei and Quinn weren't thinking about dark matter when they developed their idea. But it turns out that the properties of the axion make it a good candidate for the dark matter particle. Unfortunately, we still don't know if the Peccei-Quinn theory is merely an elegant model or a good representation of how the universe actually works. One way to check is by looking for experimental consequences of the model that we can detect. And just a few years after physicists began to explore the properties of the axion, it was given new purpose when it was realized that it had the basic properties that we think dark matter has, based on astronomical observations: potential for lots of the particles to exist and slow moving (and therefore cold). Part of what I like about the axion is this feature: we need it to solve the strong CP problem and by total coincidence, it also satisfies the requirements for a good dark matter candidate. The axion is a good dark matter candidate because we need it anyway.

Axions are also exciting because they behave differently from how we might intuitively expect the missing matter to behave and can display some interesting quantum properties. This is because axions are scalar bosons like the Higgs that I discussed in Chapter 1. Because bosons are willing to hang out together, they can display fantastical behaviors. They are called bosons in honor of Indian mathematical physicist Satyendra Nath Bose, who was one of the earliest contributors to theories of quantum mechanics and collaborated with Albert Einstein on developing the equations that describe particles with whole-number spins. A consequence of these Bose-Einstein statistics, as they are known, is the Bose-Einstein condensate (BEC), where a large number of relatively cold particles or atoms all go into the same low-energy state and proceed to act like one single super atom. Bose-Einstein condensate formation uniquely reflects the quantum mechanical nature of matter, which sometimes behaves as if it is made of billiard ball-like particles and sometimes acts as if it is sloshing like watery waves. In the BEC state, the particles not only act like waves, but the waves add together coherently to create one super wave. This is a purely quantum mechanical phenomenon. In classical non-quantum physics, the particles would act like little unbreakable billiard balls and could never add up together the way they do when they are waves. The quantum BEC phenomenon is extremely difficult to produce in the lab and not expected to happen much in the universe, except maybe in neutron stars.

Well, it wasn't expected to happen much until physicists started thinking about what axions might do if they were the dark matter. In the last two decades, it has become clear that axions are an example of a special class of dark matter that now has a series of names: scalar dark matter, fuzzy dark matter, scalar field dark matter, and wave dark matter. These are all variations on a theme: "scalar" is another word for a boson with spin zero. The Higgs is a scalar, and the fact that we've observed it gives us confidence that perhaps other scalar particles exist too. The word "field" arises because when you merge quantum mechanics with Einstein's special relativity, a mathematical structure called a field is needed to describe particles. A simple example of a scalar field is the temperature of a room: the temperature has a value at every place in the three-dimensional room. If you're using central air, the temperature is probably about the same everywhere. In the open-concept living and dining room where I am writing, we only use a fireplace in winter, so the temperature changes significantly from the couch, which is near the fireplace, to the dining table, where my feet are currently very cold. A particle is described by a similar mathematical construct, but with a hefty dose of quantum mechanics. Sometimes the field description of a particle becomes particularly meaningful. Bose-Einstein condensate dark matter is one of those times.

The fact that the axion can behave like a BEC means that there may be macroscopic quantum waves floating around in space! How big they are depends on the mass of the axion, and theory doesn't actually provide us much insight into what the mass should be. We have constraints from experiments that rule out certain values. There are also observational constraints that suggest a preferred value. In my research, I have found that at this preferred mass scale, axion BECs the size of an asteroid could have formed in the early universe. Do they survive into the present day? This is a problem that I am actively working on.

There are other axion mass scales that create very different astrophysical outcomes and tie into other areas of theoretical physics too. I've mentioned that we have been unable to fit gravity into the Standard Model. One possible way to address this is well known as string theory, which proposes to merge gravity and quantum field theory, the calculational structure behind particle physics. In exchange we have to accept that spacetime, which we traditionally think is made of three space dimensions plus one time dimension, may have at least ten spatial dimensions. String theory is incredibly elegant, and I like to joke with my friends that when you have so many dimensions to play in, anything can happen. It turns out that a lot does. For example, string theory has many by-products that include phenomena called moduli, and these moduli have very similar properties to the axion. In fact, the quantum chromodynamics (QCD) axion, as the Peccei-Quinn axion is known, is part of a larger class of particles with similar features that appear in models of quantum gravity—theories that unify gravity and quantum physics.

We are no longer dealing with just one kind of axion—QCD—but now dealing with a whole class, known as axion-like particles, or ALPs for short. With ALPs, the macroscopic quantum wave would be the size of a dark matter halo, which is the dark matter that envelopes a galaxy. The Milky Way has its own halo, and it's much bigger than the visible stars that we can see on the edge of our galaxy. If the halo is made of ALPs, the smaller structures, like the tens of galaxies that are satellites of the Milky Way, will have a distinct formation history compared to what may have happened in other dark matter models. Exactly *how* different is something I'm currently working to figure out.

And yes, there are other dark matter models. The axion-like particle is distinct from another class of proposed solutions to the missing-matter problem, weakly interacting massive particles, also known as WIMPs. Rather than being a specific particle candidate, WIMPs are a group that generally share the set of properties of interacting with each other via the weak force (one of the three forces described in the Standard Model) and being massive (often thought to be around 100 times the mass of the proton). Lots of candidates for WIMPs exist, and all of them require structure beyond the artifice of the Standard Model. Many of them fall under the umbrella of supersymmetry, an extension to the Standard Model that as yet has no experimental evidence for it, despite intense efforts by experimentalists at CERN's Large Hadron Collider.

When it comes to WIMP dark matter, we don't expect the dark matter halo to be full of potentially asteroid-sized or halo-sized coherent waves, because WIMPs are typically made of fermions. They can't form Bose-Einstein condensates and behave much more like classical particles.

Anyway, I don't like talking about WIMPs because they're not axions, and I work on axions. But it's important to note that WIMPs remain an important line of research. That said, they're not the only alternative contender to axions. For a while, a popular alternative to the WIMP was the MACHO—massive astrophysical compact halo object. You can tell that physicists love an acronym—a set of initials that spells out a word—and that the physicists who came up with WIMP and MACHO were almost certainly men. MACHOs are effectively ruled out by data, but something like them has recently become a popular candidate again.

Black holes, the strange objects predicted by general relativity where space and time switch properties so that once you're in, you can't get out, are an increasingly popular candidate because of the detection of gravitational waves. This is an old idea that got new life in the last few years. We are in a new era of black hole discovery, and what changed is the detection of gravitational waves, which are ripples in spacetime that occur when very massive objects are in motion; for example, two black holes that are gravitationally bound to one another—a binary. Gravitational waves are analogous to the ripples you see when you drop pebbles in a pond, or a fork in a sink of dirty dishwater. They were hypothesized for a long time, and we had even seen indirect evidence for them from pulsating neutron stars, but it wasn't until 2015 that they directly vibrated our detectors.

These vibrations came from the collision of two neutron stars, a special type of star that is the product of stellar death; when a star is massive enough, its death cycle begins with a supernova explosion and ends with a neutron star. These objects are extremely dense—imagine sticking two and a half suns into a space the size of Los Angeles—and are made almost entirely of neutrons, so we think some really fun physics that can be summarized as "hot quark soup" happens inside of them. They're terribly fun objects because they are basically space quantum chromodynamics laboratories, and if you feed them too much mass, they collapse into a black hole. Because they are so massive, two of them circling each other in a death spiral causes sufficiently large ripples in spacetime that we can now detect them with special instruments.

In the years since the first detection of gravitational waves from neutron star binaries, there have been detections of black hole binaries and black hole–neutron star binaries. We've learned a lot from these detections about the production of heavy elements like gold, which is made in collisions between neutron stars, and we've also learned that black holes with masses similar to stars—stellar-mass black holes, as they are known—are more common than what was predicted by theory. When experiments and observations contradict theory, things get exciting because there are new questions to ask. Increasingly physicists are wondering: What if dark matter isn't matter at all, but rather pockets of extremely distorted, gravitating spacetime? Why would the universe have so many black holes? Black hole dark matter is special for another reason: it's the only scenario where I might agree that "dark" is a good descriptor because they tend to absorb light and not give it off.

It's worth pausing to consider some of the patterns and backstories here. Physics is often represented as a linear sequence of singular (white, male) heroes making discoveries. The reality is more complicated. Sometimes theories come first. Sometimes the experimental data comes first. People in different locations can arrive at the same conclusion at essentially the same time, totally independently. Scientists often disagree on whether something is even a problem in the first place. Articulating scientific questions is social. An idea can germinate across the generations for decades before someone really figures out how to get at its innards. Kelvin, Poincaré, and Zwicky are all historicized as genius giants of their time, and mysteriously Vera Rubin generally hasn't been. In Rubin's lifetime, men younger than her won Nobel Prizes for finding evidence of phenomena that were similarly significant to finally proving the existence of something behaving like an invisible matter (the 2011 prize for dark energy). Another woman, Jocelyn Bell Burnell, watched as her male PhD adviser was awarded the prize for her observation of the first neutron star. Meanwhile, in my discussions about theory advances, Helen Quinn and Chien-Shiung Wu are the only women who make an appearance.

Today, much of my research focuses on dark matter and neutron stars, with a particular interest in axions as a dark matter candidate and neutron stars as axion laboratories. In general terms, the gendered history of these objects and my work on them is largely a coincidence. But maybe not entirely. As a graduate student who was struggling with self-confidence and her place in the theoretical physics community, I once spent a day with Dr. Rubin. When I met her, almost the first thing she did was ask me how I thought the dark matter problem could be solved. No one had ever asked my opinion about a major problem in physics before, not in grad school, and definitely not in college. Physicists are rarely interested in the opinions of undergraduates, who are often not deemed advanced enough to make a useful contribution to the conversation. In research, the convention is that one hands an undergraduate a problem to work on and hopes they will be creative in their efforts to apply known techniques to tiny fractions of difficult problems. In some sense, I understand this. And yet some of the most interesting research I have done as a scientist was a collaboration with undergraduates who dug in and went beyond their coursework, asking questions and finding interesting threads in the work. Importantly, physics is about precisely that, which means that we should probably be asking our undergraduates big-picture questions, not for the sake of getting solutions out of them, but to encourage them to take ownership over those questions.

But what if you never give yourself permission to think about a problem? This is what happened initially between me and dark matter. I understood it as a problem for observational astrophysics, not particle theory, and as a result, I wasn't particularly interested in it for myself. I thought it was an important problem, but not *my* problem. Vera Rubin changed that impression, simply by asking me the question. What Dr. Rubin did was probably more anti-establishment than it already seems. Not only did she ask a student what she thought about how to solve one of the biggest problems in physics, she also asked me, a brown-skinned woman.

I suspect that along the way some undergraduates do get asked for their opinions. I watched as a few of my white, mostly male classmates were feted as next-generation physics greats. It became clear to me early on that the faculty had decided that I was not going to be in that particular group. Why ask someone who would never amount to anything as a scientist what she thought about science? Dr. Rubin challenged how I was treated in physics by treating me as if I was a person who could solve a major problem in physics. At the time I wasn't working on dark matter and didn't have a good answer for her. It was another five years before I ended up in dark matter's intellectual orbit, and even at that point, I had no intention of becoming a person who works on dark matter. A colleague suggested we try to make sense of a proposed theory about how axion Bose-Einstein condensation works. As a result of our efforts, we produced an interesting result and suddenly I found myself, Dr. Rubin's encouragement at my back, thinking that I too could be one of the people hot on the trail of dark matter.

Today dark matter is one of the two great mysteries in cosmological physics. Occasionally scientists claim that dark matter sounds scary and foreboding to the general public. Actually, I've only ever heard white scientists make this claim, and whenever it happens I wonder if that's really telling us more about the way they relate to the idea of "dark." In my view, dark matter is extremely benign. If it was going to hurt us, it already would have. In fact, dark matter plays a critical role in making us possible: dark matter must have come into existence early on because it is essential to the formation of all structures in the universe. The other great cosmological problem, cosmic acceleration, comes later in the timeline. But only after dark matter has helped create structures like galaxies that are full of stars, some of which have solar systems, one of which is ours. Despite the role it plays in our existence, to work on understanding dark matter is an act of faith. It is a strange and thrilling topic to choose as a research area because first of all, you have to hope it's not completely invisible. If it is perfectly invisible, with no interactions whatsoever with leptons or quarks, there may never be a direct detection. We may never be able to say, "There it is, right there in the data." And we never really know from where, or from whom, the most interesting ideas about it will come.

THREE

SPACETIME ISN'T STRAIGHT

I didn't even know we had a problem with quantum gravity.

—Daniel "Desus" Baker,
Season 1, Episode 8, *Desus & Mero*

As a physics student I was told repeatedly that I intuitively experienced space and time as completely different phenomena. As a physicist I wonder if that was ever true or just physics professors projecting their experiences onto us undergrads. In reality, I don't think I ever gave it much thought before I was told what to think about it. In contemporary knowledge systems that have their roots in the eighteenth-century European Enlightenment, our sense of space and time are definitively distinct, and we are taught that space is flat. In some sense, it was Isaac Newton who codified this for Western thinkers—by treating time as if it passed uniformly from the perspective of all observers, in other words universally, and by allowing it to be a parameter through which all motion in space is interpreted. This was the beginning of modern physics as we know it, and it's still the first thing physics students learn in college.

To construct this artifice, Newton relied on a series of ideas about geometry—the mathematics of shapes and space—which has deep roots that go back to the city of Alexandria, now the second-largest city in the African nation of Egypt. The mathematician Euclid, born while Alexandria was under Greek control, wrote a text that left a lasting impression on the Mediterranean region, *Elements*. In it, he set out a concept that nearly every American student must learn in high school: Euclidean geometry. Euclid's geometry was originally based on five fundamental axioms—assumptions (although modern scholars have realized that we needed more than five to actually make the whole of Euclidean geometry work). The original five were intended to set out how to describe the mathematics of a plane, which is effectively like a flat sheet of paper. To describe a plane, the following rules were required:

1. Any two points can be connected by a straight line.
2. Any straight line can be extended into a longer straight line.
3. Given a straight line, a circle can be drawn such that the straight line is the radius of that circle.
4. All right angles—the angle between two lines that are perpendicular—measure the same, 90 degrees.
5. Two parallel lines remain parallel, and all lines that are not parallel must eventually intersect.

These rules seem abstract, but they are the basis for how intellectuals on the Asian peninsula called Europe eventually came to think about space—composed of three intersecting planes. If you have any vague recollection of grade school mathematics, you might remember these as being characterized by the x, y, and z axes. The first class I took that touched on the question of space at all was high school geometry, and very specifically Euclidean geometry. In our geometry classes, we are told that this is the way the world is, describable by these three perpendicular planes (see figure 4). Assuming Euclid's axioms, one can prove that the angles of all triangles always add up to 180 degrees.

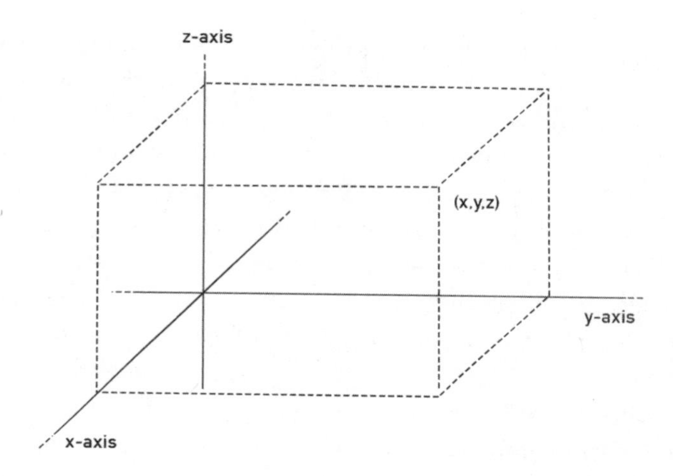

Figure 4. The three planes of space that undergird Euclidean geometry.
S. Zainab Williams

Many of us are told in high school geometry that, of course, these Euclidean geometric rules don't apply on the surface of the earth, which is curved and only seems flat at short distances. If you use large enough distances, a triangle on the earth will have angles that add up to more than 180 degrees. Euclidean geometry is a local theory—it is relevant to the length scales that we are used to in everyday life. We can't see the curvature of the earth from human heights. The roundness of the earth is not modern knowledge; by the time Euclid was born, Pythagoras had already proposed that the earth might not be flat. Aristotle agreed that it was round. Not too long after Euclid died, Eratosthenes, who was born in the Greek-occupied city of Cyrene (in the present-day African nation Libya), made the first accurate measurement of the distance around the earth, the circumference. And still, around the Mediterranean and the European peninsula, Euclidean geometry was cast as the only geometry. It wasn't until the 1800s that European mathematicians widely accepted the possibility and legitimacy of other geometries. Even in high school geometry courses today, we only learn a little bit about these geometries. Riemannian geometry (named for Bernhard Riemann), for instance, gets covered only at the end of the course, but it is seen as less important than Euclid's—even though we live on this spherical planet which cannot be adequately described in Euclidean terms.

There's a bit of an irony in the nineteenth-century refusal to accept other geometries because for most of the two millennia that preceded the nineteenth century, many astronomers and philosophers from Egypt, Persia, Greece, and Europe assumed that the space beyond Earth was shaped like a sphere. The origin of the idea of a celestial sphere actually seems pretty intuitive to me—it arose in tandem with an understanding in that Mediterranean region that the earth was not flat but in fact spherical. If the object on which we live is spherical, perhaps the sphere is a fundamental object. That, at least, is what Plato thought. Plato was wrong, but I can see why he came at things that way. I've always found it somewhat surprising that people didn't, at that point, consider the possibility that space itself might be curved. For nearly two millennia, their view of the universe was that the round earth was situated inside a flat space that was bounded by celestial spheres. Two sets of beliefs about the nature of space were held simultaneously. The surface of the earth was curved and the surface of the sky was curved, but everything in between was flat.

This celestial sphere went through many iterations over the centuries, and it was not disrupted as the dominant model of space until Newton, with some help from Johannes Kepler's laws governing planetary orbits, used the idea of a universal gravitational force to give a unifying explanation for Kepler's laws. Newton introduced an entirely new mathematics—originally known as the calculus of infinitesimals—to explain these laws. This mathematical system is so foundational to physics that when people ask me what I do all day, I say, "Calculus." Calculus is the mathematics of change while geometry is the study of shapes. In geometry, we are most concerned with straight lines, and in calculus, we are obsessed with curved ones. But Euclidean geometry didn't disappear with calculus; instead, it formed the foundation of Newton's *Principia Mathematica*, where he introduces his calculus and the natural philosophy that we now teach as "classical mechanics" to first-year physics students.

Newton's conception of objects moving in space relied intensely on Euclidean geometry as an organizing framework. In a sense, students are still introduced to calculus through this lens. Every curved line can be broken down into a series of small straight segments, so small that we call them infinitesimally small. To calculate the integral of a mathematical function, you calculate the area of infinitesimal boxes underneath the curve the function makes when you draw it. (Mathematicians eventually moved from this geometric conception to something more abstract, but physicists still often think in these terms.)

Almost everything in my education about space eventually came back to Euclidean geometry, because it was supposedly intuitive. In introductory physics, where we first learn Newton's mechanics, all our initial calculations are cast in René Descartes's Cartesian coordinates—that familiar x, y, and z—which describe locations within planes that obey Euclid's geometry. We are told that this is the best way because it is the most intuitive way of looking at the spatial organization of our world.

The Palikur people of the Amazon see it rather differently. Their geometric system, which more accurately describes the movement of stars across the night sky than the Euclidean one, is what we would call "curvilinear." Understanding stars moving across the sky requires a kind of intuition for curves—something that's hard to gain when you're always thinking in Euclidean terms. The Palikur system seems to train the mind to think in terms of curves from the very start, rather than using straight lines as a jumping-off point to curves. In my relatively rudimentary understanding of their mathematical way of looking at the world—which is based on interviews conducted by academic anthropologists Lesley J.

F. Green and David Green—Palikur geometry takes curves into account and uses anacondas as a fundamental geometric object, rather than abstract straight lines. In an interview with a visiting anthropologist, a Palikur expert on these systems explained that he felt sorry for people in other parts of the world who didn't have access to their knowledge.

Indeed, as a scientist trained in the Western tradition, I am probably an appropriate object of his pity—even though my Euclidean intuition was something that didn't come to me naturally the way it seems to for other physicists. Raising questions about the supposedly intuitive, universal Euclidean geometry also leads me to wonder about the scientific presumption that time flows forward, uniformly and independent of observers. I promise, this is not a joke about CP time. But every person has their own internal clock, and our cultural contexts around the counting of time vary. The prevailing Western wisdom is that we individually and collectively typically experience time as a one-dimensional phenomenon that always moves forward. Yet the Maya had a concept of cyclical time, and this shouldn't be so counterintuitive for someone like me. I experience menstrual cycles that are roughly the length of a lunar month. As anyone who experiences menstrual cycles will know, they are rarely identical, either; as we get older, they change. When my cycle starts, it usually feels like time has rebooted, right back to the beginning all over again. And it is not just the same moment, but in the same place. Again, my uterine lining and blood are being shed through my vagina. Yet I also know that my body is aging forward in time, my eggs are getting older, and that with each period, I get closer to my last one. After my last one, my vagina will likely change. Do we really even perceive space and time as separate when we mark time by events that occur in locations? Blood flowing out of our vaginas—the moon appearing to have a certain shape in the sky. Which sense of time is the correct one? The one that marches forward and never repeats, which seems to be organized around the universal guarantee of decay, or the one that centers and recognizes repetition?

Some might argue that there is human time the social phenomenon and then there is absolute time, the physical phenomenon measured by clocks. The essential point I hope you will take away here is that maybe *intuition* about space and time isn't universal and that it has cultural and experiential context. What constitutes intuition and our "gut" feelings about what models of the world make the most sense must therefore also be a social phenomenon. It is something we are taught, rather than something we are born with. Luckily for my apparently misaligned intuition, it turns out that the Euclidean-Newtonian way of looking at space and time, while often useful for doing calculations in certain scenarios, is in fact not entirely accurate. I will never know if I really believed that space and time were separate because of some kind of human intuition or just because the educational system I was brought up in emphasized this with regularity. What I do know now is that the separation of space and time is a lie, even from the point of view of a clock.

Like many ideas that felt intuitive for Europeans and their settler colonial descendants, this particular intuition about the separateness of space and time is fundamentally wrong. It didn't come to them because their worldview somehow made it seem natural. Physicists had to be pushed by experiments to accept this. In the late nineteenth century, the Michelson-Morley experiment showed that the speed of light was finite, suggesting that the universe has a kind of speed limit. The experiment was designed to test a longstanding hypothesis that space was filled with something called a luminous aether. This was intended to explain how light could travel throughout space because although light was already at that point understood to be a wave, it was believed that waves had to travel on a medium. An example of the root of this intuition is looking at waves in the ocean—there is no wave without the water. It was similarly believed that light must be riding something, and the luminous aether was postulated to be that something. The Michelson-Morley experiment was designed to test this idea, and it found no evidence whatsoever for the luminous aether. Contrary to the idea that scientists are strict empiricists—people who take experimental data at face value—many scientists at the time resisted this diagnosis about the properties of light and went looking for ways to rescue the hypothesized aether.

Einstein, though, wasn't one of them. He is often remembered as a genius for his insight that the speed of light implied radical new ways of looking at space and time, introducing the basis for arguing that space and time are in fact one four-dimensional phenomenon. What I like to pause and consider though is that what Einstein did best was take seriously the logical conclusion of the idea that the speed of light was constant and that light really could move through empty space as a wave. In other words, he accepted the reality that experiments like Michelson-Morley were suggesting, and it led him to develop some revolutionary physics, the first part of which we know as special relativity. Questions remain about how much assistance Einstein got in this work from his first wife, Serbian physicist Mileva Marić, and the role she played in special relativity may never be adjudicated to broad agreement among historians of physics. We do know for sure that Marić, extraordinarily, was only the second woman to graduate from the degree program where Einstein was only unusual in his relative mediocrity for a male student. We also know that at the time they graduated and got married, few women in Europe were allowed to hold professional positions doing scientific work of any kind, and married middle class and higher women were absolutely expected to confine themselves to unpaid full-time jobs as domestic laborers and managers.

Einstein was not the only person to take the results of the Michelson-Morley experiment at face value. Between 1889 and 1904, mathematicians and scientists George FitzGerald, Hendrik Lorentz, Joseph Larmor, and Henri Poincaré all developed concepts that became the cornerstones of special relativity. FitzGerald and Lorentz proposed the idea that when an object is moving, its length is shorter than when that length is measured while the object is stationary, or not moving. This is premised on the idea of frames of reference. If, for example, you're driving a car, while you're sitting in the driver's seat, relative to the rest of the car, you are stationary. You drive by someone who is standing on the street. In fact, it looks to you like the person and the street corner they are standing on are moving toward you and then as you pass them, they look like they are moving away from you. The person on the corner has a different frame of reference. They are standing still, and it is you and the car who are moving. What FitzGerald and Lorentz each separately proposed was the idea that explaining the Michelson-Morley experiment required acknowledging that the person driving the car and the person standing on the corner would disagree about measurements of length in each other's frames. The car would look shorter to the person on the ground.

In the years following Michelson-Morley, subsequent experiments that searched for evidence of the luminous aether and tested the nature of mass demanded explanations that built on the FitzGerald-Lorentz Length Contraction Hypothesis. In everyday life, we tend to think of "mass" and "weight" as meaning the same thing, but in fact, mass is a deep physical concept that is defined as the amount of an object's resistance to change in motion due to a force. This resistance amount also happens to determine the attraction between two gravitating bodies, say for example between a human body and the celestial body of Earth—a specific attraction that we know in everyday parlance as a person's weight. This attraction of course changes if we switch out the earth for the moon. On the moon, you weigh only 16.5 percent of what you weigh on Earth, but your mass stays the same. The difference in weight occurs because the moon is much less massive—contains less gravitating stuff—than the earth, so the force between you and the moon is smaller.

In the years just before Einstein's famous 1905 paper introducing special relativity, Larmor, Lorentz, and Poincaré are all responsible for extending the FitzGerald-Lorentz contraction to include not just length contraction but a parallel phenomenon involving time called time dilation. Time dilation essentially goes hand in hand with length contraction in the sense that if two different frames of reference have a relative speed, like a person on a street corner and a person in a moving car, you disagree not just about lengths but also about times. How much time passes while you're driving down the block? You and the person on the corner will disagree about this. Time will pass more slowly according to the person on the ground. The difference is so small that it usually isn't noticeable and doesn't matter. But it matters when things are going close to the speed of light, as particles like neutrinos sometimes do.

In 1905, Einstein postulated two ideas that synthesized the work of these scientists:

1. The laws of physics are the same in all *inertial* frames of reference, which means in all frames of reference where there is no change in velocity (which captures speed and direction), known as acceleration.
2. The speed of light is the same for all observers in all frames of reference, no matter how they are moving and no matter how the light source is moving.

In his 1905 paper "On the Electrodynamics of Moving Bodies," Einstein showed that assuming these two ideas led naturally to length contraction and time dilation, providing a foundational framework for what had previously been ad hoc ideas. The title might seem surprising because I haven't really said anything about electrodynamics or the related concepts of electricity and magnetism. Yet the motivations here were not simply about the Michelson-Morley experiment, but also trying to make sense of developments in the theory of electricity and magnetism that were inconsistent with what was known about Newton's physical laws. Michelson-Morley is in fact intimately connected with what we call electromagnetism. Just a decade before the experiment, James

Maxwell had completed the theory of electromagnetism, which showed that electricity and magnetism are in fact two sides of the same coin and were the source of light, which behaved as a wave.

Before I go on and explain how this leads to understanding spacetime as a unified concept, let me pause to note why I'm explaining things this way. I've told this story with a historical perspective not because I want to reproduce a historiography of great white men, but because too often when we speak to the public about physics, we make it seem like a litany of lone geniuses. Rather, physics is an intensely social phenomenon, and that has only become truer with time. The ideas that come to populate physics, especially the ones that stick, are rarely the product of one person's ideas but rather the result of a community effort. For example, in the instance of special relativity, Einstein wrote a paper that synthesized key ideas, but it was in part composed of pieces that had been constructed by others in the years before. As students, we were taught that Einstein's unification of space and time in special relativity implies that the ruler we use to measure distances in "spacetime" is described by something called the Minkowski metric. It was Hermann Minkowski, not Einstein, who came up with the metric that governed distances in Einstein's new science. The Minkowski metric is distinct from the traditional Euclidean metric because it captures that the speed of light is the same in all inertial frames.

As lovely as special relativity is, it is limited in its capacity to completely describe spacetime because it did not capture gravity. The way we tend to think about gravitation is inherited from Newton. In his picture, space is like a room that you fill with things and where actions happen. Technically speaking, we call this the background. Space sits in the background while action happens inside it. In this context, gravitation is a force—an attraction between massive objects (which is to say, most things)—that operates inside the room without ever affecting the room itself. In special relativity, this continues to be the case, although now we have some new ways of thinking about how distance and time are measured inside the room. Importantly, Newtonian gravity faces a challenge when we try to bring it into conversation with special relativity: the strength of the gravitational force depends on the distance between two objects, and as I've just explained, measuring distance depends on the observer. Making gravity and relativity work together requires reimagining gravity—and it took ten years after the unveiling of special relativity to make it work.

Einstein's key insight that brought gravity into the relativistic fold and produced the theory of general relativity was this: a person in a windowless rocket ship in space that is accelerating at the same rate that gravity pulls objects to Earth will not be able to tell if they are in a rocket ship or in an elevator that is sitting on the earth's surface (see figure 5). We also experience this briefly when we feel weightless on a roller coaster. These are examples of how physics conforms to the equivalence principle, which states that gravity is not a real force but rather a pseudoforce that can be mimicked by acceleration—which is to say a change in speed or direction that occurs at the same rate that it would under gravity. Gravity is not, in fact, real.

Figure 5. As described in the text, Gravity Girl's ball behaves the same in both scenarios!
S. Zainab Williams

The consequences of thinking through this realization about gravity and the questions it raises are enormous. To describe this mathematically required a new ruler that went beyond Minkowski's metric—one that used mid-nineteenth-century Riemannian geometry, which, it turns out, describes curved spacetimes. Gravity is not a force but rather an artefact of a curved spacetime. Imagine going down a long slide that has some curves in it. We are tied to spacetime in the same way, following the bumps as the shortest possible way to get from the top of the slide to the bottom. Riemannian geometry described a mechanism for constructing a new type of ruler that would take this into account. The source of the curvature? Matter causes spacetime to bend around it, and we experience this bending of spacetime as a force, gravity. To return to our room analogy: spacetime isn't a room that sits in the background. It is a room that is interacting with its contents. The contents can only move around the room according to its shape—as before in the Newtonian picture—but now the contents of the room also influence the shape of the room. The room is in the foreground, changing with time. For centuries, scientific thinkers had assumed that gravity was a property of matter, but in fact it was a property of spacetime.

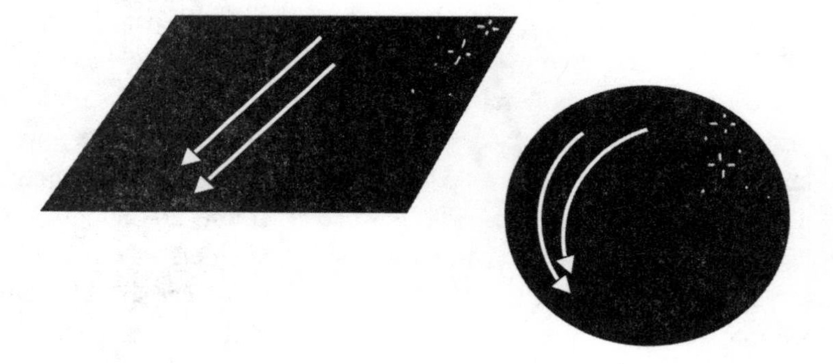

Figure 6. On a flat Euclidean plane, parallel lines will never intersect. On the surface of a sphere, two lines that start out parallel will eventually intersect.
S. Zainab Williams

You may think this has nothing to do with your everyday life, but if you've ever used a computer to get driving directions, or flown in a plane, then chances are your life has been affected by GPS, the Global Positioning System. This military technology that underpins consumer products like Apple Maps counts Gladys West, a Black woman mathematician, among the people who contributed to its development, while Roger Easton, Ivan A. Getting, and Bradford Parkinson have been traditionally credited with its invention. If GPS just relied on Newtonian physics, it would send you to the wrong location, regularly. This is because just as in special relativity, general relativity introduces time dilation. If space is curved, and space and time are linked, then time also has to be distorted—dilated. Without taking this gravitational time dilation into account, the clocks that help make GPS go would be inaccurate. General relativity is part of everyday American life, probably more so by now than apple pie ever was.

The mathematics that describes general relativity is, to steal a phrase from Carl Sagan's *Billions and Billions*, enormous gorgeous. The spacetime of this theory is smooth and continuous, meaning that no matter how closely we look at it, there are no holes. Each point is connected to another point. My senior year of college I took general relativity as a graduate course at MIT, taught by a visiting professor, Belizean American mathematician Arlie Petters. Professor Petters changed my life in a couple of ways. First, he was the first and only Black professor I've ever had, and he took a personal interest in my success, which I really needed. But second, he taught the class from a textbook that's considered extremely difficult to learn from because it is so abstract and mathematical. Robert Wald's textbook *General Relativity* was perfect for me. I loved his definition of manifolds—the mathematical spaces that are matched with physical spacetime in general relativity. Manifolds describe a space that close in looks Euclidean, but when you zoom out may look very curved. For example, here at my house in New Hampshire, the earth seems pretty flat to me. But I know that when I fly home to Los Angeles, the pilots must use a route that takes the curvature of Earth's surface into account.

Until that general relativity class, I had been convinced I wasn't smart enough to do the kind of science that required expertise in relativity. When I began studying from Wald's book, I realized I *had* to do it. What I learned in Petters's class delighted and shook me. I loved me some manifolds, I did. While "introductory" Euclidean geometry had been completely counterintuitive for me, "advanced" general relativity felt like the most natural thing in the world.

Any standard general relativity course builds up the tools necessary to derive and then calculate with what we call Einstein's equation, which describes the relationship between spacetime's curvature and spacetime's matter-energy contents. We call this a dynamical relationship because both the curvature and matter-energy create change in each other. Now I'm going to include an equation, the only one in this book:

$$G_{\mu\nu} = 8\pi T_{\mu\nu}$$

This equation looks very intimidating, but I am going to explain it and the way it creates the difficulty for us theoretical physicists. On the left side, we see the letter G, with a little subscript of Greek letters, mu and nu. On the right side, we see 8 times the transcendental number pi, which many of us may recall has approximately the value of 3.14, multiplied by a capital T, mu nu. To read this out loud I would say "G mu nu equals eight pi T mu nu." The G, which is called Einstein tensor, represents the curvature of spacetime, and the T, called the stress-energy-momentum tensor, represents all of the matter-energy content. In other words, the part of the equation that contains all of the information about the properties of spacetime is equal to the part of the equation that contains all of the information about the properties of the matter-energy content of the spacetime. This is just another way of thinking about what was discussed above: spacetime tells matter how to move, and matter tells spacetime how to curve.

I mentioned that G and T were tensors but didn't say much about what that meant. A tensor is a mathematical construct that is a convenient way to organize lots of information. The mu and nu are what we call indices, and they take on the values of 0, 1, 2, and 3, corresponding to time, *x*, *y*, and *z*, respectively. So there is a G_{00}, and a G_{23}, and this means the equation for any given component works out like $G_{23} = 8\pi T_{23}$. Because there are four possible values each for mu and nu, there are sixteen total components for the tensors G and T. Each component of the equation like G_{00} contains a part of a mathematical equation, so in effect Einstein's equation is actually sixteen mathematical equations. Because of symmetries akin to the spherical symmetry I discussed in Chapter 1, we can actually show that ten of the equations are redundant, leaving us with six. Six equations is a lot! And what's in them?

On the G-side of the equation, there is a series of operations that we call a derivative. The derivative mathematical tool was developed simultaneously by Isaac Newton and Gottfried Wilhelm Leibniz to describe change in physical systems. An everyday example of a derivative is the change in location of a car over a period of time—this change is the velocity. When you ignore the direction the car is going in, velocity is simply the speed, the number you read off the speedometer in your car. In effect, Einstein's equation is describing how the ruler we use to measure spacetime—the metric—is changing dynamically in space and time due to the presence of some matter-energy content. The exact nature of the derivatives that appear in Einstein's equation was actually developed in the decades before general relativity, by our old friend Riemann.

On the T-side of the equation, what we include depends on the matter-energy content of the spacetime. We tend to convince ourselves that most things can be described as a fluid. In big simulations of the universe, dark matter, for example, is often treated as a fluid. The equation to describe fluids in general relativity includes some velocities, the force on the surface of the fluid (also known as the pressure), and information about how much mass there is in the fluid. All of this information put together allows us to understand the structure of spacetime, given some basic information about what's in it.

In theory, that's how it works anyway. In practice, what I have described for you is a differential equation, and differential equations are (often) easy to write down but then terribly difficult to solve. Generally speaking, differential equations describe relationships between a quantity and changes to that quantity. The differential equation that we call Einstein's equation describes a relationship between the metric—the ruler we use to measure distance and the passage of time in the universe—and how that metric changes at different points in spacetime, depending on its contents. Solving Einstein's equation means finding the mathematical form of the metric. And the reason we want this metric is that this tells us exactly how spacetime is curved, assuming we know what's inside it.

For some differential equations, there is only one solution, and we know what it looks like. But many differential equations can have lots of solutions. Einstein's equation is no different. This makes sense physically—there are different conditions out there. There are parts of space that are empty, and there are parts of space that have a lot of stars or dark matter particles in them. The ruler that we use to describe the behavior of spacetime is different for empty space and for space that has a roughly spherical object in it. The ruler we use also depends on the length scale. I use a different metric for studying the spacetime around a neutron star than I use for studying all of spacetime. In both the case of the spacetime around a neutron star and looking at all of spacetime, I am able to rely on certain symmetries like those I described in Chapter 1, which is how I am able to calculate the exact form of the metric. Spacetime has special symmetries: on large scales it is homogeneous—about the same everywhere—and isotropic, meaning that at every point, almost everything looks the same no matter what direction you look in. Homogeneity and isotropy are core properties that provide the theoretical foundation for cosmology (which I describe more in the next chapter), and in the last hundred years, they have also been affirmed by observational evidence, over and over again. They also make the math simpler.

A lot of the time, we can't solve Einstein's equation exactly, but that doesn't mean we give up. There are other techniques available to us, often referred to as "numerical techniques." We have special algorithms that allow us to calculate the values a solution has, even if we don't know how to describe the solution with a succinct formula. These algorithms can be executed by hand, and in fact that is what the Black women mathematician-physicists who used to be "hidden figures" did for NASA. Today, very little of this work is done by hand because nearly any computer you can get your hands on has some capacity to do these calculations, even the potentially out-of-date iPhone you might be clinging to. What used to be a full-time job for the human computers at NASA can now be the subject of homework assignments for our undergraduates at the University of New Hampshire.

For the most part, we've come to grips with gravity. Except there are limits to our understanding. In earlier chapters, I took a deep dive into particle physics and their quantum properties. I barely mentioned gravity and didn't really discuss spacetime at all. Why? Because yes, Desus, we have a quantum gravity problem. Quantum mechanics teaches us that the smoothness of everyday life is another approximation that doesn't hold at the smallest scales. At the smallest scales, everything is quantized, meaning that it is the opposite of smooth and can be broken down into discrete parts. In this new way of looking at the microscopic world, fundamental properties of matter come in discrete quanta, like the energy levels that particles can have. For example, because they are fermions, electrons in a helium atom can't have any energy but rather can only take on specific values of energy. We believe that somehow spacetime begins with quantum properties like this, but on large scales it is what we would call completely classical: smooth, continuous, no weird quantum effects.

In my view, the deepest problem in fundamental physics is explaining just how this theory of a quantum world and our theory of spacetime can be brought together. Doing theoretical physics involves challenges of two different kinds: the conceptual and the mathematical. The mathematical problems actually always trail us around, even if we resolve the conceptual problems. Specifically, once you come up with a conceptual picture in physics, typically you can write down a mathematical equation that approximately describes the concept. But, there is no universal guarantee of the ability to solve the equation, either using brute force or simplifying symmetries. Quantum gravity, a theory that merges quantum mechanics and gravity, is, for the moment, stuck at the conceptual stage. We can't even all agree on how to think about what it should do, much less begin using mathematics to describe it.

Theoretical physicists have proposed lots of interesting ideas, some of which you may have heard of: string theory and loop quantum gravity are just two of them. String theory is by far the most widely researched proposal, and it has also captured the public's imagination. It's easy to understand why: one of its most basic proposals is that spacetime isn't just made out of three space dimensions and one time dimension, but rather there are at least ten spacetime dimensions and possibly up to twenty-six. In the string theory picture, at the smallest scales, all particles are replaced by one-dimensional quantum objects that are called strings. One of the most exciting aspects of string theory is that it promises to unify all four of the major physical forces—gravity, electromagnetism, the weak nuclear force, and the strong nuclear force—into one theory. As things stand, we currently have three of them all unified under the Standard Model, but gravity remains outside of this picture.

Despite its possibilities, string theory faces challenges, including a dependence on the extension of the Standard Model I mentioned earlier, supersymmetry. Many of us expected (at least hoped very much) to see evidence for it at the Large Hadron Collider during the last decade but we simply didn't. This doesn't mean that string theory is ruled out, just that we still don't have evidence for it. Because string theory has so many dimensions, it is mathematically rich, which makes the possibilities for it incredibly fascinating. It's also very hard to calculate in string theory, so one of the challenges we face is our ability to maneuver in it. String theory isn't the only quantum gravity model to face that particular problem, though. They all do. And if you ever want to see physicists get emotional, stick proponents of different quantum gravity models in a room and tell them to discuss the relative merits of their models. Some string theorists will smugly suggest that their theory is the best because it draws inspiration from pre-existing particle physics and shares much of its mathematical structure.

People who work on loop quantum gravity (LQG) could retort that only *their* model takes into account some of general relativity's deepest lessons about the nature of spacetime. Rather than beginning with particle physics, LQG begins with the idea that spacetime is, as I discussed above, dynamical and not static. It then tries to directly quantize general relativity with this idea in mind, leading to the concept that at the smallest possible scales, spacetime comes in discrete quantized units, or loops, that are knitted together and feel continuous on large scales. A rough analogy for this is that cotton shirts are pretty good at keeping us covered, but if you took a microscope to one, you'd see that they are made of threads woven together and there are in fact small holes in the weave. My first peer-reviewed scientific paper took this idea and tried to apply it to a cosmological problem (cosmic acceleration, discussed in the next chapter) by proposing that maybe the knots were non-locally connected to each other, rather than linked in the more intuitive side-by-side way. Perhaps the cosmos is disordered, I proposed—with ample support from my co-author and PhD adviser Lee Smolin, whose ideas, along with those of my PhD committee member Fotini Markopoulou-Kalamara, inspired me to suggest this possibility.

Unlike string theory, loop quantum gravity doesn't promise to explain the intersection of quantum mechanics and spacetime *and* particle physics, just the former. In that sense LQG is less ambitious than string theory because it only proposes to explain how spacetime is quantum mechanical in nature, without unifying gravity with the three other forces of nature. LQG is less popular than string theory and there are debates about the reasons for this. Researchers in LQG, like Smolin, would say it is because the mainstream community isn't open-minded enough. String theorists would argue that LQG is more off the beaten path from mainstream physics and less productive because it cannot explain particle physics. This is still a purely theoretical debate—there is no observational or experimental evidence for either string theory or LQG—and as yet it cannot be adjudicated with data. There's also a debate about whether that will ever be possible too.

Do we as a society have a quantum gravity problem? Desus's (hilarious, as in, I *hollered*) comment was part of a segment about the most accurate image the scientific community has ever captured of the edge of a black hole—the event horizon, which is the point of no return. Being able to get up close and personal with a black hole's event horizon may give us clues to quantum gravity because it turns out that black holes are great laboratories for thinking about the intersection of quantum mechanics and spacetime. Another way of asking the question is, Does it matter to humanity whether we know what the theory of quantum gravity is? I don't know. It's not clear that it mattered in 1910 that gravity hadn't been integrated into the theory of special relativity, but in the end, uniting the two led to interesting technological developments like GPS, and arguably, it deepened Western society's spiritual connection with the night sky. I tend to find that each person, whether they are a scientist or not, gets excited about spacetime and the fact that it's curved for different reasons. At least, almost everyone seems to be intrigued by the question. So maybe it matters for humanity because we are the total weirdos who would care.

FOUR

THE BIGGEST PICTURE THERE IS

When people ask me what I do, I say that I'm a particle cosmologist. I tell them, "I use math to figure out the history of spacetime." It's my job to fill in the details about important events on the cosmological timeline—in other words, to tell a cosmological story using math. I have work to do because we still don't have a good sense of how the story begins, although as I described in "In the Beginning," we have filled in a lot of the timeline between then and now. What I love about this work is that this is the biggest picture there is: spacetime and its (dis)contents. It also feels like being the keeper of a deeply human impulse. To borrow a word from the Indigenous communities that my Black ancestors probably come from, I am a griot of the universe—a storyteller. And although I am the first Black woman to hold a tenure-track faculty position in theoretical cosmology, I am certainly not the first Black woman to be a griot of the universe.

Indeed, every community, including those of my African ancestors, has a cosmology. My way of studying the universe—through the analytic frameworks that I have mostly inherited from the Euro-American imperialists and settlers who kidnapped my ancestors from Africa and enslaved them—is not the only one. It is not the cosmology of my maternal ancestors. It's not quite the cosmology of my paternal ancestors either. Although my father is white in most global contexts, he is also an Ashkenazi Jew of Eastern European heritage—the kind of people who come from the shtetl, the village, not urban Western Europe like Einstein. Their cosmology is the Biblical Genesis, not the general relativity that I grew up to become an expert in. Cosmology is a deeply human impulse—we have always wanted to have a sense of where we came from and why we are here. It happens to be that because of where I grew up and my own personal tastes, I study one particular perspective on this, and in the coming pages I hope to sketch out this picture for you in a way that will help you feel some of my enthusiasm for it. At the same time, I can't ignore the larger context that my work happens in, and before the chapter ends, we'll get to that too.

The way professionalized physicists understand the cosmic timeline is something like this: in the beginning, there may have been no beginning. The spacetime that humanity calls its universe may in fact be one bubble of an infinite number of bubbles, all of which may be inaccessible to us except for the one we're in. It's not clear how we will prove, with physical or observational evidence, that this is true or not true. But currently, a large swath of my professional community leans toward believing that this is true. Perhaps there was no real big bang. So, in the beginning, we are maybe just talking about our single bubble, and before that bubble was a second old and certainly before there were any Standard Model particles wandering around, it likely underwent a rapid expansion where spacetime grew faster than the speed of light—because as we currently understand things, spacetime is the only phenomenon in the universe that can break the universal speed limit. This time period is called inflation, and it was first proposed by one of my research mentors at MIT, the wonderful Alan Guth. There's much we don't understand about inflation, including what exactly set it off. It didn't last very long—less than tenths of a fraction of a

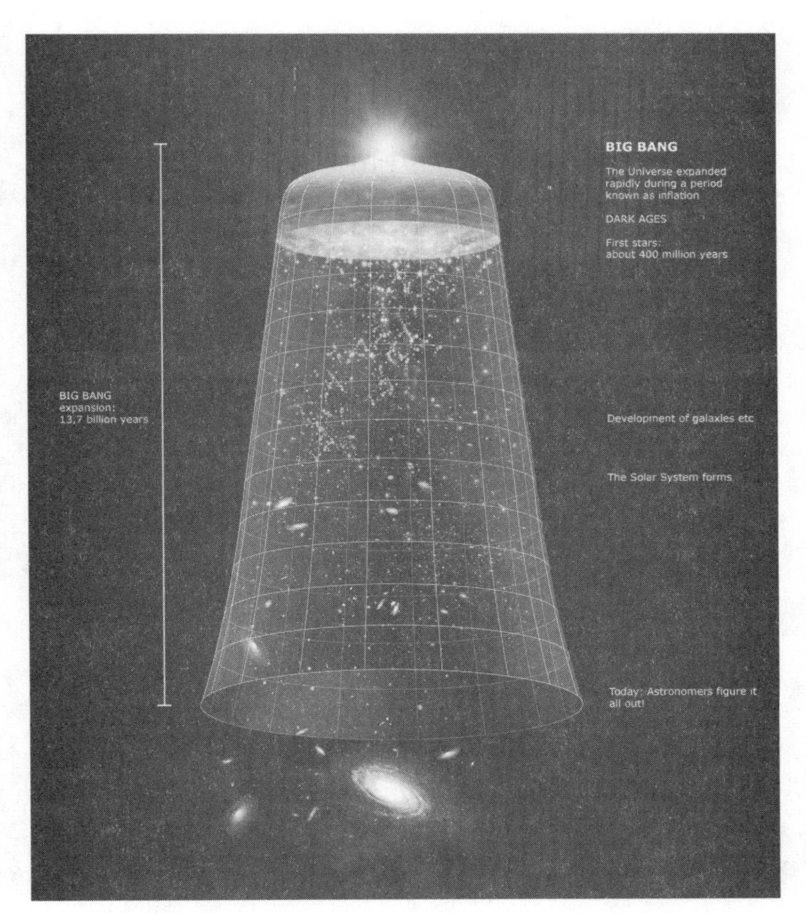

Figure 7. This is a not at all to scale history of the universe according to physical cosmology. At the top, we have whatever came before space-time existed—which may or may not have been something like a big bang—and at the bottom is the universe we see around us today, mostly empty but also containing many galaxies. It's easy to look at this image and think that we've worked it all out, but we still can't even fully simulate the formation of a galaxy, much less billions of them forming over the course of fourteen billion years.

DETLEV VAN RAVENSWAAY / SCIENCE SOURCE

second—and when it was over, the universe continued to expand, albeit much more slowly.

Inflation is a vaguely contested idea. I say vaguely because there is a small but incredibly vocal minority of people who can quite reasonably be classed as "inflation haters." In the last five years especially, there has been an uptick in the number of editorials appearing in the pages of respected publications like *Scientific American* and *Nature* that challenge the centrality of inflation to our standard cosmology, even going so far as to say it's not real science because it's not testable. Essentially, this argument hinges on the fact that while we have generally worked out what inflation does and how it does it, we still don't know any details about the particle that would cause inflation to happen, and it is easy to write down different models of inflation that all essentially accomplish the same thing. Accusations sometimes fly around in the science press and the blogosphere about how inflation is completely untestable. I always shrug a little about these kinds of objections because while inflationary theory does face real challenges, it's also the case that it's hard to explain observational data without it. It's our best possible model and that fact alone means it merits continued excavation and theoretical experimentation. As my other MIT adviser—theoretical physicist and historian David Kaiser—would probably say, the evidence for inflation also continues to be strong. But before I can explain what evidence we have, I need to get further ahead in the timeline.

Every physicist and astronomer, no matter their field, knows the basics of inflation, which is that spacetime rapidly expanded. What isn't widely discussed is what happens after the end of the inflationary era—less than a tiny fraction of a second after our spacetime first came into being. Currently it seems likely that the inflaton, the particle field that drives inflation, transferred the remains of its energy to other particles, and those particles multiplied in number. We also know that inflation should have significantly cooled down the universe. We know, too, this time from actual experimental and observational data, that in order for the rest of structure formation—galaxies, stars, us—to occur, the universe needed to reheat. But we don't know how. This question has been one of two dominant drivers behind my research in the last few years: How do you reheat a phenomenal cosmic banquet after inflation has flash-frozen it—without going too far? This is a question of general interest to cosmologists, as well as people working on technical issues in quantum field theory in curved spacetimes (particle physics at the interface with gravity). I happen to fall into both categories, so this is fun for me.

A general schematic for the expected early universe timeline is given in figure 8. On the horizontal axis we have time increasing in units of seconds from left to right. On the left vertical axis we have the temperature of the universe, increasing from the bottom to the top.

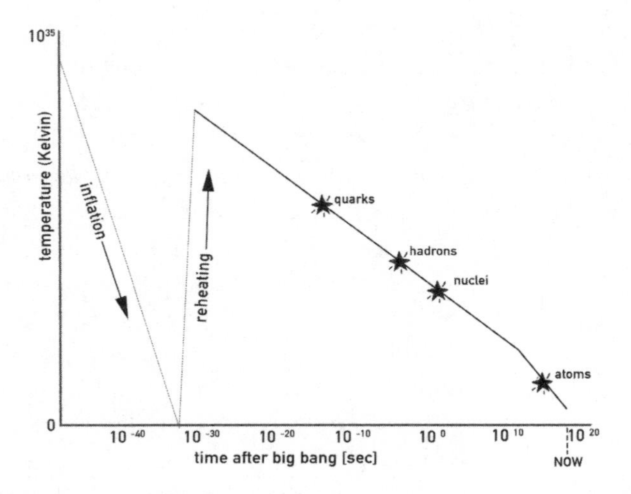

Figure 8. This is another way of looking at the cosmological timeline that you saw in figure 7, and it's much more similar to pictures I actually look at for my job; understanding how the temperature of the universe changed with time in the early universe is an important topic of active research. This particular rendering was inspired by a figure from Australian astronomer Charles Lineweaver's 2005 paper "Inflation and the Cosmic Microwave Background."
S. Zainab Williams

There's a dotted line on the plot that's labeled reheating. It's a very nice line. But it's just a placeholder because we don't actually know what it's describing or what its shape should really be. This is also the case for the inflation line, which makes the reheating problem even harder. All we know is that at the end of inflation, the particle(s) (inflatons) that drive inflation should continue to give off energy and transfer that energy to other types of matter, thereby reheating the universe, ideally so that the dotted line meets up with the solid line at a point that we should be able to predict mathematically (this is called the thermalization temperature).

So, that's what reheating is: turning the universe up a notch by energy transfer from the primordial inflaton to the matter that together will come to be known as the Standard Model of particle physics. At the very beginning of the reheating era, we expect that this energy transfer will happen efficiently through a phenomenon called parametric resonance. Anyone who has ever been on a swing has actually experienced this. Remember pumping your legs at just the right point to make the swing go higher? You were transferring energy from your body to the swing at the perfect moment to maximize the energy transfer—this is parametric resonance. The first stage of reheating, preheating, is what happens when the matter that will eventually evolve into everything we see swings higher and higher as the inflaton pumps its legs at just the right point.

In my own research on this with Matt DeCross, David Kaiser, Anirudh Prabhu, and Evangelos Sfakianakis, we looked at the aftermath of a model where there is more than one particle driving inflation (that's the "multi-field inflation"), and those particles have a deeper relationship with spacetime than just living inside of it. The idea of more than one particle shouldn't seem too radical to us at this point. After all, we used to think that protons and neutrons were fundamental particles. Then it turned out they had quarks inside. Initially we thought there were only three quark flavors. Now we know there are six. We also know that there are huge differences in mathematical and physical richness among the theories without quarks, with only a few quarks, and with all the quarks we know of. This is what we expect of the inflaton. If it is in fact a composite particle, this is consistent with how things often are in particle physics.

There is another layer of complexity in the type of model we studied. The question of how the particles interact with gravity can be addressed in one of two ways. The minimal, simplest way is that the particles interact with the metric just like everyday matter. The more complicated way, technically known as non-minimal, is via a direct relationship with the part of Einstein's equation that describes curvature. You might wonder why we would even bother with something so complicated, when there is already a simpler way of doing things. The answer is partly scientific and partly social. The social answer is that this non-minimal, multiparticle model is exciting. It turns out that models like this have a rich mathematical structure that translates into interesting physics, including what the strength of the relationship between the particle and the spacetime can be in order for parametric resonance to occur and be most efficient. The more scientific answer is that to make this theory—which lives somewhere in between particle physics and gravitational physics—explain our observations and not introduce new tensions with particle physics, a non-minimal relationship between the inflaton and spacetime might make more sense.

In either case, this is all speculative. This kind of research is at the edge of our understanding of what happened to particles and spacetime in the immediate aftermath of inflation. A lot has to happen in our hypothetical timeline after preheating ends—because somehow we have to arrive at planet Earth and the formation of humanity. That means we need atoms, and importantly, the particles that make up atoms we described in Chapter 1. Preheating gives us a good kick in the right direction, and it helps the universe reheat to a temperature—a high enough energy—where baryogenesis could occur. Baryogenesis is the process through which matter comes to dominate over antimatter, and we need it because otherwise matter and antimatter would have completely annihilated each other, leaving no possibility of structures forming—of *us* forming. We have lots of theories about why this asymmetry happened, with new proposals appearing almost daily. We also still don't know what reheating process gives us a high enough temperature (which, again, is a proxy for energy levels) to make it possible. We are fairly confident that baryogenesis happened after reheating, and when it was done, the universe began to cool. At this lowered energy, symmetries broke. First, the strong force separated from the electroweak force. After more cooling, the electroweak force separated into the discrete electromagnetic and weak forces that we are now more familiar with. This time is known as the quark epoch because it's the only time in the history of the universe (or at least our corner of it) when quarks roamed spacetime freely. As you may recall from Chapter 1, quarks are always bound together in mesons and baryons, which together are the hadrons. But this was all very short-lived—we still haven't reached a second after the maybe big bang yet.

As the universe continued to cool, quarks coalesced almost entirely into neutrons and protons. At this point, it was still too hot for electrons to stick to protons, so there still weren't any atoms. Finally, at a second old, leptons (electrons, muons, taus, and neutrinos) became the dominant matter in the universe. At this point, the universe was hot enough that lepton/antilepton pairs were created with high frequency. But this phenomenon couldn't last because the universe continued to cool. Ten seconds into this universe's existence, photons took over as the dominant form of matter in the universe. This photon epoch is when nucleosynthesis began: most of the hydrogen isotopes and helium that will ever exist in the universe were formed during the first few minutes of the universe's existence, and they began as nuclei during nucleosynthesis.

As we examine the cosmological timeline, so far we've only looked at timescales that range from fractions of a fraction of a second to a few seconds. But the photon epoch that comes after inflation exists on a different timescale and moves us into a tens-of-thousands-of-years-old universe that is completely dominated by matter—the non-photonic components of the Standard Model of physics. By 400,000 years, the universe had sufficiently cooled so that photons could no longer keep the hydrogen and helium in their ionized state as nuclei. During this era, the time period known as "recombination" occurred: the hydrogen and helium nuclei combined with free electrons to form the first long-lived neutral atoms. At the same time, the photons that were previously occupied by interactions with hydrogen, helium, lithium, and beryllium—but were no longer energetic enough for that activity—were free to stream through the universe.

Looking forward, it turns out that about 14 billion years later, a carbon-based life-form on a small planet orbiting an average star in a spiral galaxy measured the temperature of the empty space near-ish to a star and found that it had a non-zero value. The carbon-based life-form, a human as the life-form calls itself, was named Andrew McKellar, and many of the humans on the planet agreed with him that the year was 1941. Twenty-three years later, two more carbon-based life-forms of the same kind, this time called Arno Penzias and Robert Woodrow Wilson, measured a light signal with a long wavelength—in the radio bandwidth. They were not expecting this signal. As in, finding it was an accident. This is how the story was taught to me as a student: Penzias and Wilson were experimenting with a radio telescope and trying to get rid of unwanted interference when they ran into a signal that wouldn't go away and seemed to be the same no matter what direction they pointed it in. Around the same time, Robert Dicke, Jim Peebles, and David Wilkinson were setting off to find just such a signal, which they had predicted from theoretical work on the early universe. A mutual friend of both teams of scientists had seen a draft of the paper making this prediction and mentioned it to Penzias and Wilson. The radio signals they found are now understood to be the cosmic microwave background radiation, CMB for short.

McKellar is rarely credited for his 1941 measurement, which was the first evidence of a CMB. In some sense, the CMB is the first light in the universe—it is the first time in the cosmic timeline that photons were able to stream freely, and that is effectively what they continue to do today.

The CMB tells us an enormous amount about spacetime and its contents about 400,000 years into the timeline. Helpfully, we can associate a temperature with any given wavelength of light. Thus we can convert the CMB's radio waves into a temperature of about 2.73 Kelvin,* just three degrees above absolute zero. The existence of the CMB means that nothing in the universe can currently be colder than 2.73 Kelvin. As time goes on and the universe expands, the light wave will stretch with spacetime and the temperature associated with it will go down. But for now, we're at 2.73 Kelvin. In the years since its discovery, NASA has launched two telescopes into space to better understand the CMB. The first was COBE, the Cosmic Background Explorer. Between 1989 and 1993, COBE provided extensive evidence that the temperature in the sky was indeed almost entirely uniformly about 2.73 K—on large scales, the universe is indeed homogeneous (each spot looks the same as any other spot) and isotropic (everything looks the same in all directions). But it also provided evidence that on smaller scales, there are what we call anisotropies, or variations, in the temperature.

It is here that I can make good on my promise to explain some of the evidence for inflation. The large-scale homogeneity paired with small variations in temperature, and the way they get imprinted on the CMB, are best explained by the inflationary model. As I described earlier, during inflation, spacetime expanded rapidly. In fact, it expanded faster than the speed of light. This meant that parts of spacetime that have not since been in causal contact with each other still have similar properties, and it answers the

* Kelvin is a unit you get when you add 273 to the degrees in units of Celsius.

question of why parts of spacetime that are far apart from each other "know" to have these similar properties. This homogeneity and isotropy were, for a long time, key assumptions of our cosmological picture, but without any explanation. Inflation provided a reason why. At the same time, we know the universe isn't exactly the same in all locations. Some spots have lots of matter in them—galaxies—and some places are empty voids.

These anisotropies can now be traced back to quantum fluctuations of the quantum inflaton fields—the excitations of which are the inflaton particles I mentioned earlier, the ones that may have driven preheating. What this means is that particles flicker in and out of empty spacetime, and some of them stick around. This is one of the key insights of quantum physics: that particles have lifetimes, that particles decay, and also that particles spontaneously come to life. In the early universe, these flickerings translate into quantum fluctuations that show up as anisotropies in the CMB. Subsequent experiments like NASA's Wilkinson Microwave Anisotropy Probe (popularly known as WMAP) and the European Space Agency Planck telescope confirmed for us that these little variations are real, and they actually fit a relatively vanilla picture of inflation. In other words, little variations in the CMB temperature (typically one part in 100,000) are effectively a snapshot of the quantum fluctuations that began at the beginning of time and would eventually turn into everything visible in the universe, including us.

But the CMB is still only 400,000 years into the cosmic timeline—we still haven't arrived at us yet. We're still at the point where there are no stars and the universe is relatively dark. This era in the universe's history, often referred to as "the dark ages" lasted for hundreds of millions, maybe a billion, years. But eventually gravity successfully caused hydrogen and helium to clump together compactly enough that they formed stars, objects that are so dense that nuclear chain reactions begin. The seeds of these first steps toward structure formation originated in the inflationary era as small fluctuations in the empty vacuum, and they left imprints on the CMB. From the point of view of gravity, the vacuum is empty, but quantum mechanics tells us that particles will occasionally randomly flicker into existence, a phenomenon known as quantum fluctuations. These particles are the beginning of everything to come later, including the first stars. The nuclear chain reactions in these stars produced photons that light up the universe in many frequencies, including those frequencies that the human eye was later tuned to, which is what we call the visible.

Stars are giant balls of gas plus plasma—one of the four phases of matter along with solid, liquid, and gas. That means they are atomic elements plus collections of electrons and electrically charged nuclei, the part of an atom that is made of protons and neutrons. The first stars were formed out of the leftovers of recombination, and they come from a neat balancing act between local pockets of curved spacetime and the tendency of spacetime to spread things out through its expansion. On large scales, spacetime expands continuously, generally growing the distances between specks of gas and dust. But it is sprinkled with the aftermath of little quantum hiccups—little matter field fluctuations that flickered randomly into existence during the inflationary era. Because spacetime tells matter how to move and matter tells spacetime how to curve, these little quantum hiccups create tiny little gravitational wells, little dents in spacetime where matter clusters together. Over time, atoms fall into the gravitational wells, making them deeper and more forceful because the mass of the atoms strengthens the gravitational hold of the well. It is within these clusters of atoms that the first stars were born.

Before they become glowing balls of continuous nuclear explosions, stars start out as collections of gas. Eventually, the gravitational pull of the atomic gas is so powerful that the gas becomes very dense, and that creates interactions between the atoms that go beyond gravitation. The weak force and strong force become dominant, causing nuclear reactions to begin. The ignition of nuclear reactions is what lights up the stars. In stars around the mass of our sun, the reaction is relatively simple. It begins with two protons fusing together, and eventually four hydrogen nuclei decay into a helium nucleus, two positrons (the electron antiparticle), and two electron neutrinos.

This proton-proton chain reaction is happening right now, and it is the source of the light that we see when the sun rises, the same sunlight that illuminates the moon at night. More than four million tons of matter are converted to energy every single second. This process is also the source of electron neutrinos that were first detected in the 1990s by the Gallium Experiment, GALLEX, deep below the Gran Sasso mountain in Italy. In stars slightly more massive than the sun, just one and a third times larger, the nuclear chain reaction is more complex, involving carbon, nitrogen, and oxygen. Humans may recognize these three elements as critical for the formation of life on the collection of rocky materials that we call planet Earth. Importantly, stars aren't just a thing that formed in the past. They are forming in the now where I write, and the now where you read. To see a picture of a stellar nursery, see figure 9. But you'll want to see this image in color, so when you get a chance, type "Hubble Pillars of Creation" into an internet search engine. A small consumer telescope can also see a blurry version of the home nebula of the Pillars, the Eagle Nebula, if you have access to a dark sky at the right time of year for your hemisphere.

How can we possibly know so much about stars? Part of it is from looking at them and observing the frequencies of light where they are bright and the frequencies of light where they are dim. The electromagnetic spectrum is the full range of frequencies that light has. These frequencies are most familiar to us humans in the form of the spectrum of colors that many, but not all of us, can see with our eyes. What the human eye can potentially detect is actually a small part of the electromagnetic spectrum, which also includes the microwaves that we use to rapidly heat our food; the X-rays we use to visualize the bones beneath our layers of skin, muscle, fat,

Figure 9. The Pillars of Creation are truly magnificent. They are about seven thousand light-years away, which means that every photon our telescopes capture from it is seven thousand years old. The Pillars themselves are about six light-years across, meaning it would take light six years to travel across them. By contrast, it takes light eight minutes to get to Earth from the sun.
NASA, ESA AND THE HUBBLE HERITAGE TEAM (STScI/AURA)

and blood; and the cancer-causing UV rays from which our precious ozone layer and, importantly, our melanin protect us. This practice of breaking the light down into different frequencies is called spectroscopy. When the sun comes out during a rainstorm, spectra (the plural of spectrum) are a common sight: we call these rainbows. In the case of a rainy-day rainbow, the raindrops are the spectrograph, the machine that breaks the light into its different energetic frequencies and makes them separately visible. Scientists build machines that do this same work, and this is maybe my favorite mechanism for looking at the universe.

Looking at the frequency of light that we observe, or the frequency of light that is missing from a spectrum, can tell us what the radiating source is made of. That tells us a lot about what elements are present in different stars. It turns out that due to quantum mechanics, every chemical in the periodic table of elements has certain correspondences with a particular part of the electromagnetic spectrum. For example, when we electrify sodium, it emits light with a characteristic wavelength of 589 nanometers, what many of us would call "orange." Those of us who grew up in areas with orange street lamps have witnessed this phenomenon over and over. These are the sodium lamps that are now being phased out in favor of more energy-efficient (but frankly, ugly) white LED lights. In reverse, if we point light with this wavelength at sodium, the sodium will absorb it and change quantum energy states, changing its capacity for chemical reactions.

Our ability to interpret light like this is relatively new and depends on an understanding of quantum mechanics. Elmer Imes, who studied at University of Michigan and in 1918 became the second Black American to earn a PhD in physics, played a major role in affirming the correctness of quantum mechanics by measuring the spectrum of hydrogen chloride, hydrogen bromide, and hydrogen fluoride molecules. The measurements captured features of the molecular spectrum that only made sense if quantum physics is correct. These molecules are more complicated than the atoms I was describing earlier. Hydrogen chloride is a combination of a hydrogen atom and a chlorine atom, and molecules display features we don't see in atoms. As I've mentioned before, atoms display a quantum rotational behavior that we call spin. Molecules inherit this. In addition, because the atoms of a molecule are connected, they can show vibrational behaviors, akin to a string on a guitar. Both the spin and vibration impact the spectral emissions of a molecule, and here, classical physics and quantum physics make very different predictions. In the early twentieth century there was conflict over experimental results, which contradicted each other. More precision was needed to figure out why some experiments observed a smeared-out spectrum and some saw fine lines. As Black physicist and historian Ronald Mickens points out in his scholarship, Imes's high-precision experiments were the first in human history to measure the infrared spectrum of these molecules and in the process make measurements that distinguished between the spin and vibrational modes that were expected from quantum mechanics.

The infrared spectrum is a set of wavelengths just outside of what is visible to the human eye, and night goggles often work by detecting infrared light that is radiated by warm objects, like human bodies. This correlation between temperature and the wavelength of the light is a quantum feature and cannot be completely explained by the physical theories that precede the twentieth-century advent of quantum mechanics. Moreover, infrared emission and absorption is also associated with spin and vibration modes in molecules, which is why Imes chose to make his measurements of the molecular behavior in that range of wavelengths. Importantly, stars and the dust in outer space that sometimes provide a nursery for stars also radiate and absorb infrared light. In other words, spectroscopy—the study of spectra—in the lab is where our understanding of physics and chemistry on Earth helps us interpret our knowledge of nuclear physics in the stars. Imes's work was some of the earliest work in astrochemistry. Between nuclear experiments on Earth and mathematical explanations of nuclear physics, we have been able to interpret observations of stars that are very distant from us using physics that is relatively familiar to us. Looking at distant stars, which, because of the finite speed of light are necessarily much older than our star, is how we know about the composition of stars much older than the sun.

Gravity worked on those first-generation stars, pulling them together into galaxies—or as we are now sophisticated enough to know, stars end up in the grip of spatial curvature. This is what led to the formation of galaxies. Those galaxies merged to form larger galaxies. And stars can't live forever. The first-generation stars died, producing the seeds of new generations, as well as the fascinating afterlife states, neutron stars and white dwarfs. These afterlives are a kind of star in their own right: stellar ghosts are just another kind of amazing phenomenon. Neutron stars are the afterlife of stars about ten times as massive as our sun. When these big stars run out of hydrogen fuel, they supernova, fantastically blowing off their surface layers, in the process taking out any solar system that might have been hanging around in orbit. While this explosion leaves much destruction in its wake, it also leaves behind a core as dense as an atom, made up mostly of neutrons. These neutron stars are a stellar species in their own right. Some of them produce enormous magnetic fields that, because of the stars' high rotation rate, emit light that can be visible to telescopes here on Earth. They were first noticed in 1967 by Jocelyn Bell Burnell—when she was still just a graduate student!—and her PhD adviser Antony Hewish. In 1974, Hewish was awarded a Nobel Prize for their work and Burnell was not.

Someone who was paying attention during "I ♥ Quarks" might be quick to point out that the Pauli exclusion principle also applies to electrons. Indeed it does, and this is what enables electrons to counteract the gravitational collapse of the white dwarf. White dwarfs are fascinating in their own way because a major difference between neutrons and electrons, of course, is that electrons are charged particles, while neutrons get their name from being charge neutral. Neutrons don't electromagnetically repel each other, while electrons do. As a result, white dwarfs are only stable because they are made not entirely out of electrons, but rather of something we call electron degenerate matter. Electron degenerate matter begins life as a plasma. In white dwarfs, this plasma is a combination of electrons and nuclei, which are the core parts of atoms that contain all of their protons and neutrons. White dwarfs are visible to us not because of nuclear explosions but because they are very hot, and the heat dissipates as radiation. (Humans also give off light associated with our body temperatures, but we are rather cold, so the photons are all in a part of the electromagnetic spectrum that is not visible to the human eye.)

Freddie Mercury was right when he questioned whether it's even worth living forever. Normal stars can't make gold. Only dying stars, and the neutron stars that form the afterlives of some stars, can make gold. In other words, star death—including neutron star death—is an important part of the elemental life cycle in the universe. The gold formed in the deaths of other stars became part of the dust that Earth formed out of. But actually, gold isn't the best thing to come out of dead stars, although it is pretty neat, and I do like wearing it. Even though stars are the source of nearly everything important on Earth—like carbon, nitrogen, and oxygen—hands down the best thing that they do is make neutron stars. I will fight people over this. From the point of view of particle physics, neutron stars are a natural conclusion of "things we know about physics." The Pauli exclusion principle, remember, tells us that fermionic particles like neutrons can't all go into the same state. There is a pressure associated with them. In practice this means that if the right astrophysical conditions are met, an object made entirely out of neutrons can form. This inevitability might lead one to conclude that neutron stars are actually not terribly fascinating objects when in fact they probably rival black holes in the nonexistent (you heard about it here first!) hierarchy of "cool stuff that gets made in spacetime."

Stars of all kinds are, relatively speaking, small-scale stuff. The length around a neutron star is about the length of Los Angeles County. Scale matters in the universe, and while on small scales things are collapsing, on large scales they are expanding. All this while, the universe is continuing to expand. This too is due to gravity. We face a universe with dueling gravitational effects.

Around the time that structure formation really took off, around 400 million years after inflation, so did the expansion of spacetime. It started accelerating again. This is known as the cosmic acceleration problem, and we don't know why it's happening, why it's happening now, or whether or not the acceleration will change with time. The hope is that planned observations in the 2020s and 2030s will provide some answers to these questions. Cosmology requires multigenerational patience.

The question of the cause of cosmic acceleration was, for a long time, a motivating scientific mystery for me. To appreciate the problem, we have to consider how cosmological distances are measured. Effectively, we need a really big ruler. Our first known rulers, also called a "standard candle," are Cepheid variables. Cepheid variables are stars whose brightness fluctuates because of pulsations in their atmospheres. In the early twentieth century, Henrietta Swan Leavitt, a deaf woman and calculator at Harvard College Observatory, figured out that there was a correlation between the timing of these pulsations and how much light the stars emit. When we know the absolute brightness of a star, we can calculate its distance using basic physics. Cepheid variables became key in our earliest understanding of spacetime at large distances, and were the basis for Edwin Hubble's 1929 observation that the universe was expanding.

Until I was 16 years old, scientists thought that the universe's expansion rate was constant or potentially decelerating. To check this, astronomers were using Type Ia supernovae, exploding stars. As with biological decay here on Earth, what happens during and after a star's death in space is just as fascinating as what happens during its lifetime. Type Ia supernovae are explosions of stars in very particular systems: they are the end of life of a white dwarf binary—a white dwarf that is in a gravitational hole with another star, possibly another white dwarf. In other words, Type Ias are the afterlife of a star's afterlife. What makes them especially interesting as standard candles is their very consistent light curves—the line tracing how bright they are as a function of time. The light curve of a Type Ia has a very distinct shape that is effectively the same for all Type Ias. This sameness allows us to compare brightness between objects, and knowing the distance to one, we can then extrapolate from the brightness of the others how far away they are. The first papers measuring cosmic distances using this method were published in 1998, and the results shocked the world.

The story as I heard it during a 2011 observational astronomy trip to an observatory in Chile was that then graduate student Adam Riess was so focused on looking for deceleration that at first he struggled to realize what the data was telling him. He was examining the distances they calculated from their supernova observations and couldn't figure out why the rate at which the distances between the objects was changing—the change in speed of the expansion of spacetime between them—and didn't look like it was slowing down. Someone suggested that maybe it wasn't slowing down; maybe the speed was picking up—accelerating. That is how the team that Riess was helping to lead realized that the expansion of spacetime was accelerating.

Perhaps because I came of age just as the cosmic acceleration problem was uncovered, it became a driving question for me as a student. The summer before my sophomore year of college, I wrote to Riess's PhD adviser Robert Kirshner and said that I found his research interesting and would like to do my mandatory federal work-study by researching supernovae too. Kirshner was quite nice about it and offered me a job. For the next two academic years, I paid for my textbooks, tampons, and late-night mozzarella fries by preparing images of supernovae that had just been discovered by the Hubble telescope for analysis. When I started graduate school, I stayed up late trying to think of ways to solve the cosmic acceleration problem. I had no idea what I was doing at first, but eventually I found my way to a dissertation on the topic. My PhD thesis, "Cosmic Acceleration as Quantum Gravity Phenomenology," includes work on the edge of research in a theory that unifies quantum mechanics and general relativity, loop quantum gravity. Although it's been a decade since I worked on cosmic acceleration, I remain firm in my belief that it is perhaps one of our first hints at quantum gravity, and I think a successful quantum gravity theory will be one that can explain the cause of cosmic acceleration.

How are we so certain of all of this? It is a combination of the particle physics and bent spacetime that I described earlier. It is a faith in the universality of mathematical principles. It is an incredible display of consistency across the sky and in our laboratories. It is a framework rooted in the ideas of African, Middle Eastern, and Greek thinkers who lived in the ancient Greek empire and European thinkers like Nicolaus Copernicus, Johannes Kepler, and Isaac Newton. Although the question of whether Copernicus and earlier thinkers got certain mathematical ideas from India continues to be the subject of debate among historians, this intellectual framework is largely understood to be the child of the European Enlightenment and Scientific Revolution that followed on their heels. What goes undiscussed is that it is not the only way to understand the origins of the world—or how the enlightenment and revolution were paid for at all.

The cosmological work that I do is in part a product of settler colonialism, a form of colonialism that seeks to not only control territory but also replace its Indigenous population. If that seems difficult to accept, consider this: I am in part a product of settler colonialism. As I will discuss more in the next chapter, "The Physics of Melanin," my pale appearance is not merely due to my father's genetic makeup, but also my mother's. Growing up into a Black woman who is also a particle cosmologist has required grappling with all of the threads of history that bring me to where I am. How do we have all of the knowledge that we have? Why are there so many men in my history of twentieth-century cosmology and almost no women?

Throughout my career as a PhD scientist, I have clung to Henrietta Leavitt and Cecilia Payne-Gaposchkin as two of the only women greats in the history of cosmology. Leavitt, the human computer I mentioned earlier, noticed a pattern in the data that established how to measure cosmic distances. The stars she saw patterns in, those Cepheid variables, are important calibrators for the Type Ia supernovae I described earlier. Payne-Gaposchkin came a generation after Leavitt and earned a PhD at the observatory. Her thesis argued that stars were primarily comprised of helium and hydrogen. A famous male astronomer initially told Payne-Gaposchkin that her work couldn't possibly be correct and convinced her to draw a different conclusion in her dissertation. The same man's research later showed her ideas *were* correct, and while he acknowledged her ideas in his own publication on it, because his conclusion was published and citable while hers was not, for a while, he was more widely recognized than her for this knowledge about stars. You'll remember the other woman astronomer who figures heavily in cosmology history, Vera Rubin, from the chapter on dark matter. But otherwise, the cosmological tale I just told you is also historically a tale of great white men, and it is hard to understand it as my intellectual heritage, even if I have made it into my intellectual future.

Part of the problem is that the ways people who aren't white men contributed to science are often hidden. For example, when we talk about astronomy on Maunakea* in Hawai'i (and I'll be doing a lot of that in the last phase of this book), we often talk about how the seeing on the mountain is good. By that we mean there is less atmosphere at the high altitude, which means that it's easier to get a clear picture of the sky without atmospheric interference. We know about this particular location not because of European or American exploration but rather because Native

* Maunakea is also known as Mauna Kea and Mauna a Wākea. After consulting with Native Hawaiian cultural knowledge holders, I have chosen Maunakea as the convention for this book. My text should *not* be seen as the reference point for an appropriate convention. For that, readers should consult the writing of Native Hawaiian cultural knowledge holders.

Hawaiians had for centuries known it as a place to observe their own cosmology, including a connection with the father of the sky, *Wākea*. Native Hawaiians, also known as *kānaka maoli* or *kānaka ʻōiwi* in their language, *ʻŌlelo Hawaiʻi*, had a cosmology that was partly created through observing the sky at what British astronomers came to agree were valuable observing locations, for example, the top of the Maunakea volcano. In the end, I see the continuous use of unceded Hawaiian sacred spaces for Euro-American science without the permission of kānaka ʻōiwi as an example of using Indigenous knowledge to produce science without crediting Indigenous knowledge: kānaka knew the seeing was good on the Mauna. When kānaka maoli welcomed European and American guests into their lands, they also shared information about their culture, history, and geography—including about their pristine view of the night sky.

In the interim, a lot of what we know about cosmology and astronomy in general comes from the thirteen telescope facilities that have been built on Maunakea in the last seventy years as well as from facilities built elsewhere, for instance on Haleakalā on Maui. Thus our cosmology is already, in some sense, partly created from kanaka knowledge. Around the world, there are many stories of Indigenous knowledge collected by European and American visitors and then taken back to the imperial homelands to be synthesized into what we call scientific knowledge. The modern practice of botany and ecology is full of stories like this.

One commonality in all of the stories of colonialism's relationship with indigeneity is the disconnection of Indigenous knowledge from its larger cosmology. For kānaka maoli, this includes the land itself, and not just the things in it. In 2016, Hawaiian studies professor Jon Osorio described this connection in an interview on NPR: "The relationship to this land is deeply connected, it is familial, and it incurs powerful kinds of obligations and responsibilities—kuleana we call them—and also a sense of real dependence in the way that children depend on parents and grandparents." To have the land settled without permission and subsequently desecrated is an attack on a family member. European and American astronomers have long insisted that kānaka maoli should let go of this way of looking at the world, and too often they frame disagreements about the use of the land as science versus religion, which is a polite way of saying "modern versus primitive." The wisdom behind that insistence is lost on me, even as someone trained in their scientific traditions. Every time I think about it, I wonder how different things might be going on a global scale right now if scientists from the settler colonial states had always understood that the land and its ecosystems were part of the family. What if non-kanaka scientists had a scientific view that reflected this spiritual connection? Maybe we wouldn't be facing the catastrophe that is global warming.

Even as I write extensively about Maunakea, I am aware that it will always be easier for me to write about kanaka maoli contexts than it is for me to write about the cosmology of my African ancestors. As a Black person in the Atlantic diaspora, I am repeatedly forced to confront the fact that finding a place for myself in the world is permanently entangled with the legacy of slavery. Slavery colonized the bodies of my ancestors, and the memory of who they were and what their cosmologies were has largely been erased through extreme acts of violence. I am left to reckon with the knowledge that this cosmology, the one with a genealogy that is mostly European and deeply in debt to some of the same men who facilitated the torture of my ancestors, is now my cosmology too.

PHASE 2

PHYSICS AND THE CHOSEN FEW

The universe now has a few humans in a very tiny spot called Earth—and the humans are curious about how it works—but ultimately some of them are kind of controlling about what gets studied, how, and who gets to participate.

FIVE

THE PHYSICS OF MELANIN

Readers, our future is now.

—Ytasha Womack, *Afrofuturism: The World of Black Sci-Fi and Fantasy Culture*

For most of my life, all I knew about melanin is that it gives color to people's hair and skin, and it is a natural sunblock. As a physicist, I was certain that what made it fascinating was its interaction with light—what it absorbed as a photoprotectant and what it reflected as a color that is visible to the human eye. Melanin isn't just highly pigmented; it also exhibits what we call broad ultraviolet (UV) band absorption. This means that just like sodium eating a photon and becoming more energetic, melanin absorbs ultraviolet photons across a broad range of frequencies. In doing so, it protects whatever is underneath the melanin from the harmful effects of those UV rays. For these reasons alone, melanin is indeed an interesting physical system to study, both from an evolutionary perspective and from the point of view of biophysics.

But one day I wondered: what is the physics of my skin? I had never asked this question, partly because I arrogantly assumed I knew the most interesting answer, partly because my social conditions encouraged me not to think of it. As I dove into what we know about melanin, I realized that what I *thought* was interesting about melanin might be one of its least interesting features because it turns out that understanding melanin might help us build better, greener technologies. Melanin is an amazing material. Why had I not truly appreciated this before and sought to understand it? Because I know that what police officers and vigilantes think they see is a matter of life and death.

What they think they see is race, a social construction that was never biological and was never rooted in some law that governs the way the universe works. By now a lot of us have heard this, and many—though notably not all—of us have accepted it. Even if we believe that race is a social construction, it can also seem intuitive that many people define race by skin color. It feels obvious: most of us can clearly see the difference between pale skin and dark skin. But the reason we are so attuned to this particular difference and have a tendency to group people together based on it isn't required by a well-tested scientific law. The history of chattel slavery and everything that happened in its wake works to make it intuitive for us. Race was invented hundreds of years before the nineteenth-century discovery of melanocytes, the cells that produce the pigment responsible for variation in human skin tones and hair colors: melanin. Melanin, whose name comes from the Greek word *melas*, "black, dark," is found in most living creatures, and when it is studied scientifically, researchers usually use the ink of *Sepia officinalis*, the common cuttlefish. Sorting people by skin color is arbitrary for scientific purposes, as we have learned from studying DNA and biochemistry. Race, then, tells us more about how we organize ourselves than about any absolute scientific truth.

Race is a system that structures power. As evolutionary biologist, anthropologist, and sociologist Shay-Akil McLean puts it in his essay "Social Constructions, Historical Grounds," "The utility of race is its stabilization of colonial and imperial power amassed through historical and ongoing displacement and dispossession in an expanding processual global fashion." Importantly, race is inseparable from racism, and as racism and the power structures it serves have shifted with time, so has race. Those of us who call ourselves Black today have in the past been disparately referred to as negro, mulatto, quadroon, and colored. We also have, as a collection of communities and families, responded to chattel slavery—colonialism inscribed on the body through kidnapping, entrapment, and rape—by building community, theorizing identity, and developing a sense of diaspora. In 2021, the amount of melanin in one's skin is not the only reason for identification or racialization as Black. Today, Blackness is recognized as a cultural identity that is entangled with a history rooted in melanin content but not limited to it. It is an identity that many proudly claim, myself included. Black people come in shades that span indigo-blue undertoned dark brown to varying degrees of beige to what folks sometimes jokingly say is "damn near white." Black identity both is and is not about melanin—it is a social discourse. Black identity is a historical, sociogeographic construct with a real but tenuous connection to genetics.

So, there is race, which is social, and then there is melanin, which is physical and actually wonderfully fascinating. Physically, melanin exists beyond the human world in a variety of species as a set of visibly colorful biomolecules that we think are synthesized by enzymes. In humans, melanin or the lack thereof is genetically encoded. You're born with a constitutive skin color that is determined by your unique genetic makeup—which can have a complex relationship to that of your ancestors. You can add to this base constitutive layer through exposure to UV light—either the sun or an artificial source like a tanning bed. Short UV rays work superficially in the upper layers of the skin to stimulate melanin production. Longer UV rays travel more deeply into the skin and can actually cause melanin to brown. Of course, your melanin production may not be able to keep up with the intensity of light you are exposed to, in which case you burn, a possibility no matter your skin color but especially likely if your natural equilibrium state is low to no melanin to start with. In the long term, if you don't have a good base layer, you're also at higher risk for skin cancer due to UV exposure.

There are three types of melanin: the most common, eumelanin, comes in the colors black and brown and occurs in skin and hair; the less abundant pheomelanin is on the yellow-to-red spectrum and the cause of red hair and pink lips; and neuromelanin, which appears in high concentrations in the human brain, but the function of which is still an open question. And for all the hand-wringing over the years about the connection between melanin and the content of one's character, for the most part, it seems, there's much we still don't understand about melanin. Yet, despite our inability to comprehend it, the social consequences of melanin are understood intimately by many of us. Those of us who have more of this eumelanin in our skin are more likely to die at the hands of police. How intuitively someone understands this probably depends on how much they resemble Terence Crutcher, shot by police on the side of the road, or Aiyana Stanley-Jones, shot by police on her living room couch. For centuries, scientists who had low eumelanin content in their skin interpreted high content to scientifically equate to a lower intellect. Ironically, neuromelanin is abundant only in animals with what humans would call "high intelligence," and I'm willing to believe that people who have low eumelanin skin content (also known as pale people) have it in equal abundance as those who have skin that is notably richer in eumelanin.

Unfortunately, modern history is replete with examples of low-eumelanin people using high melanin skin content as an excuse to treat people badly. Sociologist and science studies theorist Ruha Benjamin notes in her book *Race After Technology: Abolitionist Tools for the New Jim Code* that race is a social technology, not simply an individual sense of self or melanin count. In one study, a dark-skinned woman was racialized differently by observers when the same image of her was altered to lighten her skin, and the results were heightened when her nose was made smaller. Other studies show that stereotypes about nose shape, hair texture, and hair melanin content function as cues in tandem with skin melanin.

These struggles with categorization began as a chaotic mix of hatred, cruelty, greed, and perversity, and they are key to understanding why until now, maybe you've never read anything that told you that melanin, the physical phenomenon, is fascinating. What we know about melanin, and what we even *want* to know about melanin requires some historical context. In a classic example of the illogical nature of racial construction and its inseparability from racism, Thomas Jefferson, who owned his enslaved Black mistress Sally Hemings* and their children, waxes on about whiteness in *Notes on the State of Virginia*: "Are not the fine mixtures of red and white, the expressions of every passion by greater or less suffusions of color in the one [whites], preferable to that eternal monotony, which reigns in the countenances, that immovable veil of black, which covers all the emotions of the other race?" In other words, the still highly esteemed founding father of the United States preferred the expressive faces of free white people to the stoic faces of enslaved Black people, and he believed these apparent differences were due to race, not relative states of freedom and captivity. There is no science underpinning Jefferson's feelings, only prejudice—not terribly different from when singer John Mayer announced, "My dick is sort of like a white supremacist." At least Mayer was honest enough to tell the truth about his preference.

Mayer's white supremacist dick reflects his cultural inheritance as a white man who finds melanin to be grotesque and antithetical to a concept of beauty. And he's not alone: one of my earliest memories of my first year in college was a conversation with a fellow math nerd—a white man—about his dating preferences. I

* I believe Hemings was coerced by slavery into this relationship, and I also believe Hemings was a person who exercised what agency was allowed in her life. Given her circumstances, I believe she made choices regarding her relationship to Jefferson, and I also understand her as an enslaved person whose choices were limited by the brutality and legal reality of chattel slavery. Contemporary scholarship continues to debate the contours of Hemings's relationship to Jefferson and the very idea of consent under those conditions.

wasn't particularly interested in him, but he nonetheless shared with me that he would never date a Black woman because he simply had not been taught to think of us as attractive. He was simultaneously conscious of his bias and completely unconcerned by it. A few months later, I helped him ask out the white woman who is now his wife—so it's clear that he still believed in my capacity to understand what a woman might want, even if he didn't fully believe in my womanhood because of the melanin in my skin.

Mayer and my college classmate exemplify how these historical conceptions of Blackness and beauty continue to play out in modern culture, as Black feminist scholar Janell Hobson explores in her book *Venus in the Dark: Blackness and Beauty in Popular Culture*. In a chapter on the representation of Black women in film, she notes that "black women can indeed be seen as beautiful—albeit with blue-green eyes and long, fair hair." She asks, "Can we envision a non- essentializing blackness that encompasses skin color variations while also disrupting white supremacy that upholds light-skin privilege?" Colorism is a conversation we still struggle to have.

I know from personal experience that I experience less racism because of colorism, and the phenomenon of colorism means that I am often the target of a fundamentally anti-Black fetish. A few years ago, I recorded an interview with Ebro Darden, a morning DJ on New York's Hot 97, a popular radio station targeting what is called in the media "urban" audiences (that's code for Black and Brown people who like hip-hop and R & B). When the video went live on YouTube, one of the first comments was about how my light skin made me a good candidate for making "good, light-skinned" babies. I was horrified, not only because of the objectification, but also because of the suggestion that my attractiveness was entangled with my capacity to be a vessel for erasing melanin from family lineage.

In her book, Hobson describes how Black filmmaker Ngozi Onwurah, a woman of biracial and binational parentage, represents the impetus underlying this comment in her short *Coffee Colored Children*.

Onwurah represents "her siblings and herself attempting to 'wash' off the blackness of their skins through their uses of soap, bleaches, acidic fluids, and cleansing liquids. This reference to whiteness as cleanliness and blackness as 'bodily waste' is implicit in this particularly jarring ritual of self-hatred." The power dynamics that were institutionalized by slave owners like Jefferson, of treating lighter-skinned slaves as somewhat more human than those who are darker, continue to haunt the ability of Black people to relate to one another.

Of course, Jefferson's white supremacy clearly didn't inhibit his sexual interest in women who had at least *some* visible melanin in their skin, and we know from the spectrum of brown that defines Black people across the Americas that many other white men certainly saw Black women as tools for satisfying their sexual desires. Clearly, Jefferson was conflicted on melanin, both finding sexual beauty in it while downplaying its capacity for a deeper kind of emotional beauty. It's also clear that his views took the form of an early version of colorism. His light-skinned enslaved captive and mistress Sally Hemings was in fact his wife's half-sister, the product of a rape inflicted by Mrs. Jefferson's father on Ms. Hemings's Black mother. Jefferson evidently had an easier time humanizing his enslaved mistress than other enslaved people whom he perceived to have faces made of an "immovable veil of Black."

Someone being dangerously generous might attempt to convince us to be forgiving of those who associated a lack of melanin with inherent goodness. In the late eighteenth century, the source of melanin in the body was unknown. The ancient Greeks thought its presence in the skin was sun and heat induced. It wasn't until 1819 that Giosuè Sangiovanni first isolated melanocytes in the cells of cuttlefish, and 1837 was the year that Friedrich Henle connected pigment in the cells of the skin with pigment in the cells of the eye. Jefferson didn't know, you might say. There are two problems with this argument. The first is that it is predicated on accepting that Jefferson's conclusion about Black facial expressions had a logic to it that isn't underpinned by believing that Black people are subhuman. Clearly, the rational conclusion is that captivity and forced labor makes a person unhappy, but of course, to come to that conclusion, you have to believe that the being in front of you is a person.

The second problem: Jefferson and his contemporaries had the scientific knowledge to come to a different conclusion. According to Dutch dermatologist Dr. Wiete Westerhof, about a hundred years before Thomas Jefferson lived, Jean Riolan the Younger "separated the skin of a black 'Aethiopian' subject into two layers." In the same year that Westerhof released his history of the human melanocyte, 2006, Drs. George Millington and Nick Levell described the encounter a little differently. Citing Riolan's own description in his 1618 text *Anthropologia*, they note that Riolan "observed . . . the blistered skin of a black subject." It's worth pausing to consider the language chosen to describe Riolan's activities. In Westerhof's description, it's not clear whether the Black person is alive or dead and whether they consented to the separation of their skin. Millington and Levell choose a word that implies no contact ("observed")—a scenario that is hard to believe given how thin skin is. In both cases, the descriptions of the event seem to struggle with properly reckoning with what happened: a European scientist treated a Black person as a scientific curiosity. In the process, Riolan found that the epidermis—the skin—could be separated into a "black" upper layer and a pale white lower layer.

Thus, as early as the 1600s, it was known that dark pigmentation was likely embedded in the skin itself, not a projection due to environment or anything related to the functions of the mind. But a curious feature of enlightened Europe was that their obsession with conquering everything was in tension with a desire to know how things worked: so how things worked had to be consistent with justifying abominable behavior. It's clear that even if Jefferson were aware of Riolan's conclusions, he would have interpreted them to mean that the black layer had trapped the white layer, or some other nonsense that justified his continued financial gain at the expense of those he kept enslaved. This was precisely what other powerful Europeans and colonial Americans did. As the Enlightenment witnessed a transfer of power away from the church to science, science became the new foundation that racists used to argue that Blacks lived in the conditions they did—not because of any moral failings on the part of the perpetrators of slavery—but because it was the logical, natural order of things.

But none of this is natural, and none of it has to do with melanin, the physical thing, either. The tradition of racism among white scientists is perhaps not surprising when we recognize that science and society co-construct one another, and there is perhaps never a more salient site for this lesson than on skin with high melanin content. Indeed, it was a supposedly enlightened Europe that enshrined the animus toward darker-skinned people in its colonial satellites as a tool to help maintain a unidirectional flow of resources, from the colonies to the heart of empire. Rather than skeptically considering the substance of these colonial sensibilities, scientists largely sought to substantiate them through a search for their scientific foundations. Science thus became a process in which bias was consecrated by scientists. Racism was axiomatic, rather than a belief requiring skeptical investigation.

Because of this, as a Black child from a biracial family growing up in Latinx eastern Los Angeles (aka East LA), I understood that racist conditions existed and that police would shoot on sight people of my skin tone and darker. Our late family friend, the former Black Panther Michael Zinzun, lost his eye to a police beating and had to sue an entire city to get some semblance of justice for it. This is simply the logic of the thing, our society. This is a structure built on logic. That logic is both nonsensical and powerful in many ways, not only because of the formidable structures of white supremacy that are so heavily fortified by it but also because of the ways it limits our imaginations. It was only 13 years after getting a bachelor's degree in physics, astronomy, and astrophysics, 11 years after getting a master's in astrophysics, and six years after getting a doctorate in theoretical physics that I first asked myself: what is the physics behind melanin, that thing that has made all of this storytelling possible?

Realizing that artificial social structures have conditioned you not to ask basic and rather obvious questions is a harsh moment for a scientist who has been trained to ask how the world works. When I asked my doctor why I was low on vitamin D, and he explained that it is harder for darker-skinned people to get adequate amounts of sun in order to produce it, it didn't occur to me to ask about the physics behind that statement. I looked into the medical side first and discovered that current vitamin D testing regimes might not even be good measures for people who have non-negligible levels of melanin in their skin. I didn't dig deeper, didn't think to apply my knowledge of physics to the composition of my own body. I didn't ask, for example, about what it meant for light to interact with melanin and what exactly the vitamin D–skin color connection is, physically.

In the last chapter, I talked about the importance of spectra in observational astronomy and cosmology. My fascination with this dynamic between atoms and light started early in high school and continued through my junior-year quantum mechanics course in university, where I learned how to calculate their properties. In fact, I was so excited by spectra that I wrote my junior thesis on using them to study the properties of extrasolar planetary atmospheres and spent the following summer building lasers, which are effectively just amplified spectral emissions in one wavelength. The fact that I have spent most of my life really excited about the interaction of light and matter is part of why I find it so strange that I had never before been curious about the specific interaction of light with human skin. And that's a really fascinating phenomenon. Recall that atoms can absorb photons of light. When we see a red lipstick, that is because the chemicals in the lipstick absorb the other colors in white light, which includes the entire visible spectrum. The red lipstick only reflects the red colors back to us.

So when we think about what it means to see melanated skin, we are seeing the color that isn't absorbed by the person's skin.

Of course, brown isn't a color that appears in any rainbow, so why is it that eumelanin looks brown to us? The color brown is made in a bunch of different ways. It can be produced by combining red, yellow, and black, as in the CMYK color model. In the RGB color model, it is a combination of red and green. Black and orange paints can also be mixed together to produce brown. Melanin's primary interaction with the visible spectrum is to absorb some of it and reflect the rest of it back. What color we see depends on the extent to which different colors in the visible spectrum are absorbed. To achieve the browns that we see when we look at the diverse skin colors that Black people come in requires diverse levels of melanin. The more melanin, the more of the color spectrum is absorbed. But, as we know, some of the darkest-skinned people have blue undertones. This means that some blue light is not absorbed by their melanin but rather is reflected back.

This begs the question: what is it in my skin that absorbs and emits light such that I am this color, this shade of brown that is on a spectrum of racialized Blackness? In the era of #BlackLivesMatter, what could possibly relate physics to the unique contours of Black children's lives more than talking about the marvelous physics of their skin color? When these questions first occurred to me, the reason why I had never considered them before seemed obvious: it was outside my area of expertise. But with time, I realized that it's a bit odd that this never came up in my coursework, although perhaps not entirely surprising. These days there's lots of talk about diversity in science, or STEM (science, technology, engineering, and mathematics) as we broadly call it; there was certainly less emphasis back when I was in high school and college.

But today, as then, when scientists *do* talk about diversity in science, it is often in terms of the untapped resource that people of color represent. The National Science Foundation is charged with improving participation of underrepresented groups in science as a matter of national security. Ensuring that the United States has a sufficient homegrown STEM workforce requires acknowledging that demographic changes are coming: in just a few decades this will be a majority-minority country, with no single ethno-racial group making up the majority. Thus, the story goes, if underrepresented minorities—typically defined as African Americans, Hispanics,* Native Americans, and Pacific Islanders—are not included in the STEM workforce in numbers proportional to their percentage of the population, the United States is at risk of being dependent on foreigners for its technological workforce.

In other words, like my enslaved ancestors, Black people in the twenty-first century—and other so-called minorities—in science are constructed as a commodity for nation building. Certainly that comparison has its limits, since most of us are more free than the ancestors who were kept in captivity—though mass incarceration disproportionately impacts people of color, especially the most melanated people in that umbrella, Black Americans. But the underlying logic is familiar: none of this is about what society can do for people of color so much as what service people of color can provide to the national establishment. In this sense, it's not surprising that no one thought to talk to me about melanin, the wonderful biomolecule that historically was used as an excuse to mistreat my ancestors. But having that thought would require believing that physics was for Black people too, rather than that Black people, to the extent that we are welcomed in physics, exist to secure nationalist power.

There are practical considerations too; returning to Ruha Benjamin's *Race After Technology*, there's a very mundane example that is also telling about our unwillingness to think about skin color variation and light absorption. Benjamin describes the way

---

* This word is often used in official literature, not Latino/a or Latinx. This means that people who are both white and Hispanic count as minorities, and rumors abound about people from Spain calling themselves a minority for the purposes of benefiting from programs for underrepresented minorities.

automated soap dispensers work through infrared light that detects the presence of a hand underneath them. In 2017, Nigerian Facebook employee Chukwuemeka Afigbo posted a video on social media of the way a soap dispenser responded to and dispensed soap to the hand of a white person. A Black person was denied the soap. It had not occurred to the technologists behind the infrared detector and the dispenser that because darker skin has a broader spectrum of absorption, the skin would absorb the infrared light rather than reflect it back to the sensor. The dark skin was invisible, not because it wasn't there, but because the detector hadn't been designed with dark skin in mind. Just as Black people disappear from view when new, life-improving technologies are being developed, so too do questions that might specifically interest us. Part of my awakening as a Black scientist was realizing that I could use my scientific skills to understand my own Blackness: physicists are trained with a specific way of looking at the world, and it is expected that we can adapt our toolkit to any problem.

When I finally began asking questions about melanin, I approached it like a scientist, by searching the scientific literature. I was surprised to learn that melanin is becoming an active topic in biophysics after a few centuries of what I would call halfhearted investigation by racists. Now people want to understand how melanin works as a material and what utility it might have for our future technologies. What might have been different if, for example, they hadn't been distracted for about half a century by the pseudoscience of eugenics, which was considered foundational until the Holocaust helped some understand its social implications? "Eugenics" is another word with Greek roots, formed from *eu* which means "good" and *genos* which means "race." And that captures the intentions of the field: choosing who lives or dies or ever even comes to exist at all using rules developed by people with the power to make the rules. Although eugenicist ideas predate the rise of white supremacy, eugenics has been a cornerstone of science under white supremacy that even today continues to thrive in some corners of science.

Although openly eugenicist ideas are considered to be somewhat fringe, it is still clear from anthropological studies of science that the biology of the disempowered—such as people with XX chromosomes (often identified as women) of any race—is less likely to be studied. When I learned that melanin had finally become a popular topic not because it was realized that people of African descent are interesting in our own right but because so many white people are getting skin cancer, I was pained but not surprised. After centuries of kidnapping, locking up, beating, raping, robbing, and killing people in large part on the basis of the human eye's perception of skin melanin content, studying the mechanical nature of melanin—its interactions with light and its movement and production in the body—became interesting only when it seemed necessary for enhancing the survival of people who don't have a lot of it in their skin. This parallels the violent experiments of nineteenth-century gynecologist J. Marion Sims, who took an interest in Black women's reproductive systems only because of what they would reveal about how to medically treat white women. Sims is still celebrated in some quarters, even though he effectively tortured Black women by cutting them open without anesthesia, arguing that we could not feel pain. Melanin isn't the first site on Black bodies that became curious when it served the health of non-Black people.

But the substance used to justify so much death may also be the key to the future of materials technology, as in what the stuff around us like phones and power lines are made of. As described in Australian university graduate Clare Giacomantonio's 2005 undergraduate thesis, melanin is an unusual conductor of electricity. Unlike everyday conductors, melanin is bioorganic in nature (as part of a living organism, humans), and it is disordered, which means that the molecule has a disorganized crystalline structure. Because of this, the electrons house themselves in different quantum mechanical energy levels in ways that are more disorganized than simpler types of conductors. These flexible properties make it difficult to study melanin with usual biochemistry techniques, such as spectroscopy. What's more, sometimes melanin behaves like a mediator of electricity and sometimes it insulates against electricity. These properties appear to be determined by the amount of water present in the melanin biomolecule—the conductivity seems to be water-activated, although there is still debate in the literature about whether this is the best explanation of its flexible properties.

While the potential health applications of what we learn about melanin are compelling, a deeper understanding of melanin could be technologically transformative. Since melanin appears to be a relatively simple disordered conductor, it potentially provides an opportunity to decode difficult concepts about this class of materials. This class includes superconductors, materials that allow electricity to flow through them without resistance. Mercury and lead, under the right conditions, can become superconductors. Implementing superconductors at large scales would minimize the loss of electricity when it is delivered from its source to households around the world, thus reducing the amount of energy required for societies to function. This sounds exciting, of course, but existing superconductors typically work only at temperatures much colder than what we find naturally on Earth, requiring helium to help things supercool. Our sun's stellar ancestors only endowed our planet with so much helium, and helium is produced in the core at a limited rate. Plus, maintaining the helium at supercool temperatures across long distances would be energy intensive—counterproductive to the goal of reducing our energy load. This means that introducing superconductors into widespread societal use requires a technological capacity that we do not yet have. Melanin could change that. Melanin may hold the key to delivering our green-energy future efficiently, if we make an ethical commitment to using the technology for the greater good (and sadly such commitments aren't yet a normal part of scientific tradition).

Of course, we may be wrong about melanin's structure, and perhaps it cannot provide insight into superconductors at all. It may instead be what is called an electronic-ionic hybrid conductor, where the key actors are not just electrons but also ions. In the superconducting model, melanin's inherent conductivity properties change dramatically when water is brought into contact with the molecule. The addition of water enhances the number of charged particles that are freely moving around, i.e., conducting. A bioorganic electronic-ionic hybrid conductor introduces other fascinating technological possibilities. Melanin may play a key role in the design of interfaces that will enable the use of bioelectronic devices, machines that integrate with human bodies—which in turn might change the landscape for disabled people like me. As a gadget geek, I find this particular possibility thrilling. As a student of human history and the occasional reader of science fiction, it also represents a new ethical challenge because of the terrifying dystopian possibilities associated with bioelectronic devices. I love the idea of being able to use a computer to download a book into my brain—until I think about the possibilities for surveillance and control.

What all this means is that melanin, the material that eugenicists argued caused people of African descent to be inherently inferior, is also the stuff of Afrofuturist techno-dreams. There is a very real possibility that our future energy distribution mechanisms, which will help minimize damaging carbon emissions by reducing waste in transporting energy from source to use site, will exist because of the melanated people Janelle Monáe called "Arch-Android orchestrated." It may be that the key to a future where we live more harmoniously with our ecosystem is written into the genetic code of Black people. June Jordan did always tell us that "we are the ones we have been waiting for."

I worry that in pointing this out I've also identified another capitalist use for Black people. It's hard to escape the logic of slavery in a society that was built on it. As I looked for information about melanin, I found peer-reviewed scientific articles that referred to outdated phenotypic classification systems that make no sense, scientifically. I honestly hadn't realized that "Mongoloid" had never made its exit from our vocabulary. I also found no research by Black scientists. When I have run across work by Black biologists that specifically focuses on Black people's genetics, it is typically in the deeply underfunded field of hematology, where people study sickle cell anemia. Sickle cell, which involves unusually shaped blood cells and can significantly shorten the lives of those who have them, does not occur exclusively in people of African heritage, but it primarily impacts Black people. The work that researchers on this subject do is incredibly important, and we actually need more people working in the field. At the same time, it's terrible that Black people in America always seem to find ourselves in emergency mode, trying to save our own lives, rather than getting to indulge our impulses for curiosity and imagination.

As we begin the third decade of the twenty-first century, we find ourselves in a moment when melanin is being treated like a useful tool, or maybe even an art piece, yet again divorced from the people who have made it most visible. In fact, after an early iteration of this chapter was published as an essay in the Winter 2017 issue of *Bitch Magazine*, an MIT professor proposed making a building out of melanin. Professor Neri Oxman's Totems project not only co-opts an Indigenous word and concept—the English translation of the Ojibwe word *ototeman*, which is the possessive description of an object that represents a family or clan in the Ojibwe community—but it proposes to use melanin as a futurist environmentally friendly material, without ever engaging with the people who have suffered for being partly made out of it. The project itself is a series of three-dimensional, rectangular boxes filled with melanin of different colors. The melanin in this case is produced from mushroom enzymes combined with amino acids, bird feathers, and cuttlefish ink.

The pieces were produced by Oxman's group at MIT's Mediated Matter Group, which at the time of this writing had no Black students, postdoctoral researchers, or staff (except, in all likelihood, the people who clean their workspaces). In describing the pieces, the group writes, "Our use of the word [Totem] is rooted in admiration and respect of and for all things alive materially and immaterially—and for the wisdom of the Ojibwe people, who coined the word, as well as other First Nations peoples, who, unlike us, saw and felt and connected with this synergy." My heart sank when I read about this; in fact, it was demoralizing. I worried that the essay that this chapter is based on had inspired the work, and was reminded of the importance of being conscious of how our proposals might be used once they are let loose into the wider world. I did not mean to propose that scientists should consider ideas like buildings literally shrouded in something akin to the skin of Black folks. I had hoped to make the case instead that the underexplored realm of melanin science pointed to a need for science to stop rejecting Black people's humanity, including the ways that Black scientists can shape actual science. What I really wanted everyone to understand is that Black thoughts, like Black lives, matter.

In a different scenario, Black people are not a material product that can be stripped for parts but more like the Marvel comic book universe character Princess Shuri of Wakanda. She's a Black woman who likes tinkering with things and spends enough time both tinkering and imagining that she comes up with wonderful new inventions and new ways of seeing the world. In a completely different context, melanin-coated buildings sound cool, like something Black and Brown people would think of out of a deep understanding of our skin and a deep pride in its technological reach. This is substantively different from white people thinking of new ways to use our bodies. I'm not the only Black scientist who identified with Shuri when the film *Black Panther* came out—so many of us spent our whole childhoods dreaming of becoming her, only to realize that in a white supremacist society, it feels impossible. Shuri is what happens when Indigenous intellectual curiosity is not stifled. America is what happens when it is.

For the most part, what we got instead is what is often referred to as "race science." A more honest nomenclature would be "racist science." Calling it race science, as we often do, makes it sound like it was just a scientific hypothesis about race. Rather, it was a whole artifice invented to justify the superiority of (white) Europeans, their fear of the people who were different from them, and their fear of admitting that the deity they believed in could have a more expansive vision for the world than one ruled by ruthless eumelaninless men from a small Asian peninsula. The consequences of this worldview were devastating and involved both orderly destruction and disordered, unpredictable outcomes. Africans were brutally kidnapped, held captive, enslaved, forced to build families and make new generations in captivity, required to endure in new languages—and we did.

What Africans and their descendants in the Americas lived through varied greatly and the extent of our suffering has in part depended on how much eumelanin our DNA is encoded to produce in our skin. The encoding varies for reasons that are not at all random. We know that the majority of Black Americans have some European heritage, mostly due to rape, and certainly this must be common across the Americas. People often assume that my light skin is simply because my father is a white man, and indeed, he is a white Ashkenazi Jew. I am light-skinned because of who my dad is, yes, but I am also light-skinned because of who my foremothers' rapists were. My mother is not nearly as light as me, but she is also on the lighter end of the spectrum. Our melanin—and our lack thereof—tells stories about what my ancestors endured.

Though I refuse to define our ancestors solely through their experiences with violence, it is true that this violence has left its mark and shaped our experiences. Growing up, I watched my mother experience racism, and as her child, this became part of my earliest encounters with the world. As I grew older, I became more conscious of my own experiences with racism and their impact on me, even as colorism literally lightens my load compared to many Black people. For people with far more melanin in their skin than me, or even my mother, the viciousness of structural racism is more ever-present in daily life. The incidents are more frequent, more heightened, and more likely to be dangerous. For these reasons, it's tempting to define Blackness as a response to the question, "How badly do white people treat you?" It's also tempting to define Blackness as coming from suffering, as living through suffering, as the source of suffering. It is tempting to make "Blackness" and "suffering" and "melanin" synonyms, as if that is all we are.

But as a community that was forced to construct from fragments in the harshest physical and psychological conditions, we have been intensely creative. Our ancestors never stopped imagining us as a free people, and by holding fiercely onto this imagined Black future, they ensured that Black people like me would one day be able to look at the stars, not because we are on the run from slave catchers, not because we are trapped in fields, but because we learned in our first semester as PhD students in astronomy about how stars are just perpetual nuclear explosions. This is a Black magic that as a scientist I believe in: the ancestors made sure that we got here, to now, where there are new possibilities, where we are now able to ask the question, "What is the physics of melanin?" We now live in a world where a Black scientist can ask that question and understand in technical detail how knowing the answer could radically change the world.

SIX

BLACK PEOPLE ARE LUMINOUS MATTER

> What is it about the subject matter of quantum physics that it inspires all the right questions, brings the key issues to the fore, promotes open-mindedness and inquisitiveness, and yet when we gather round to learn its wisdom, the response that we get almost inevitably seems to miss the mark? One is almost tempted to hypothesize an uncertainty relation of sorts that represents a necessary trade-off between relevance and understanding. But this is precisely the kind of analogical thinking that has so often produced unsatisfactory understandings of the relevant issues.
>
> —Karen Barad, *Meeting the Universe Halfway: Quantum Physics and the Entanglement of Matter and Meaning*

BLACK PEOPLE ARE NORMAL, IN CASE YOU WERE WONDERING. I don't mean to say that we are not extraordinary in our resilience, extraordinary in our global cultural contributions, extraordinary in our survival and thrival. We are these things because we have had to be. But we are also not actually Magical Negroes. We are human like white people, like Asian people. Despite an apparently common belief, we cannot convert a cell phone into a gun. And we do feel pain. And we shouldn't need to be extraordinary to be seen as people deserving of life. Yet, Black mediocrity is never acceptable in American society. Narratives about Black people are framed around deficiencies so that to be an average Black person is to be a personal failure—and not only that, but to also fail the entire race. Black people are treated like we are perpetually abnormal relative to white people: we are a permanently failing people.

Of course Black folks have not quietly accepted this story. We have been extraordinary in our survival. We have been deeply creative in our resistance to white supremacist tall tales about us. This comes with complications, though. For example, the dark matter analogy. Because I spent a lot of my academic career in a physics-oriented bubble, I was late to the game in finding out that it was not unusual for folks in African American/Africana/Black studies to analogize between Black people's existence and the dark matter that seems to pervade our galaxies. The first time I heard about it was actually because a white physicist had jumped at this use for an article in a major newspaper in October 2015. As I said to my friends after reading it: "I never ever want to see a white scientist writing an article comparing Black people to dark matter *ever* again. I never ever want to see a white scientist compare galaxy formation to colonialism *ever again*." Later, after I saw this happening in the writings of Black people and Indigenous people, among others, I tweeted things like, "If you send me a thing comparing Black people to dark matter, I will mutter under my breath about you. It's a terrible fucking comparison."

I feel strongly about this analogy as a Black person who is a dark matter theorist, but I also wanted to think through the origin of the "dark matter: Black people" analogy specifically because of the significance it seems to hold for so many Black writers and thinkers whom I both know and respect. In the foundational collection *Dark Matter: A Century of Speculative Fiction from the African Diaspora*, released in 2000, editor Sheree R. Thomas introduces the text by explaining how dark matter and Black speculative/science fiction (sf) writers are analogical. She correctly highlights that from the point of view of physics, "most of the universe's matter does not radiate—it provides no glow or light that we can detect." Thomas goes on to write, "Like dark matter, the contributions of black writers to the sf genre have not been directly observed or fully explored. For the most part, literary scholars and critics have limited their research largely to examinations of work by authors Samuel R. Delany and Octavia E. Butler, the two leading black writers in the genre." On the latter point things have changed: N.K. Jemisin, a Black woman sf writer, is now a record-breaking award winner who has definitively entered the literary mainstream.

Obviously the enormous success of a third Black speculative fiction writer is not the reason I have an objection to Thomas's justification for the analogy. Instead, I don't agree with her proposal that exploration of dark matter and Black sf writers are comparable phenomena. In the case of dark matter, it's not completely clear why we don't understand it. It just hasn't worked out for us so far. This is partly because its very nature makes it fundamentally difficult to interact with. It is distinct in the universe specifically because it is elusive. We know little about it, but that is not for lack of trying. Indeed, we have spent millions of dollars trying to understand its nature, and we have sent some of our best trained and highly prized minds out into the theoretical and experimental world, hoping to learn more about it. By contrast, the hidden nature of Black writers is an entirely manufactured and not at all natural problem. There has traditionally been little effort to directly engage with the work of Black sf writers. The moment she went looking for them, as Thomas chronicles in her introduction, there the writers were, making themselves visible.

It's worth unpacking the larger question of Black visibility. What, exactly, about Black people is invisible? And does the apparent invisibility of Black sf writers extend more generally to Black people? After all, Black people are, often due to melanin, hypervisible: we are colorful, not invisible. Still, Black people do have to contend with a certain kind of erasure, specifically the erasure of our role in our own labor and the significance of our lives as human beings. The value of Black existence is diminished by a white supremacy that is globally pervasive. Black people are denied ownership over our own labor—whether that means the historical theft of labor through kidnapping and slavery, or the creative work that has undergirded so much of American cultural production. Because our ownership is this denied, the value and fact of our labor *as* our labor is rendered invisible.

In some sense, Thomas's essay suffers from this very problem. In the place in her words where I might expect her to ask where Black scientists are—to acknowledge that she neither mentions nor cites the work of any—she does not. Black physicists and our specific labor addressing the dark matter problem become invisible by virtue of being a non-thought in the text. Fifteen years after Thomas's essay, a white woman theoretical physicist, Harvard professor Lisa Randall, used this concept, analogizing between people of color and dark matter as two invisible forces that hold the world together in an editorial for the *Boston Globe*. Randall credited the idea to an artist that she does not name. This suggests she thought it was his idea and that she did not investigate the idea's origins before publishing about it, producing a kind of erasure of Thomas as a key interlocutor in the concept's development.

Over and over we see a common thread: humans render each other invisible and sometimes actively hide each other from plain sight. No such force is expected to be at work with dark matter. But my problem with the dark matter analogy goes deeper. Importantly, this analogy has served to misinform white people, including white scientists, about Black people—or to support their sometimes deeply held, active misunderstandings. Dark matter doesn't reflect light. It can't be illuminated. I suppose I will agree that white people sometimes think this of Black people, but that does not mean they are correct. We know almost nothing about dark matter, but we know a lot about Black people. Even white people know a lot about Black people. And most importantly,

Black people are made of the same stuff as white people, resulting in the same levels of humanity. At worst, you are saying that we are invisible and exotic rather than very visible and very human.

The comparison may be poetic but it reproduces the problem that the analogy exists to highlight. If you feel the urge to compare Black people to dark matter, *resist*. There's a lot of dark matter passing through you right now. The density of dark matter is 1 proton mass per cubic centimeter, so a few per coffee cup. There are not a lot of Black people of any size passing through you, promise. There's approximately five thousand cubic centimeters of blood volume in each human. So a lot of dark matter particles for all that blood. Why is the dark matter passing through you? Because unlike Black people, it doesn't interact with other matter normally. There's also this: if Black people are so invisible, how come the police are so good at shooting and beating our friends and family? Part of the point of Ralph Ellison's novel *Invisible Man* is that Black people are extremely visible—what's not visible is our humanity. That's not because of how we are built; that's because of how white supremacy structures our social relations.

If dark matter as an analogy doesn't serve us in understanding anti-Black racism, perhaps the way we think about the universe can still provide a way to step outside of our usual language and think abstractly about our society. We must have care in making this kind of move; there is always the potential for misrepresentation. Because Black people are extremely underrepresented in physics compared to a discipline like Black studies, there's a real possibility that a misunderstanding could propagate widely without disruption. The starting point might be to avoid looking at differences between people, and instead focus on the system that implies extensive differences. Rather than analogizing Black people, we can think about what might be a good metaphor for racism itself. There is one to be had in physics: weak gravitational lensing.

Light has to obey the contours of spacetime. Like an ant walking on a table, it can only traverse the trajectories that constitute the shape of the table. It cannot walk on a surface that is not there. I often like to summarize this in my talks by saying that matter shapes spacetime and spacetime tells matter how to move. One of the reasons I became so enthralled with relativity as a student is because this neat dance of feedback between spacetime and the things that live on its four-dimensional surfaces is neatly encoded in the equation that describes general relativity. As I discussed in Chapter 3, Einstein's equation has two sides. On one side is curvature, then an equal sign, and the other side has the matter-energy content. In effect, you can read right off the equation the grand lesson of general relativity: matter and curvature are entwined.

This means that if spacetime is curved, light, which we know always takes the shortest distance possible, travels on curved paths. It can only follow the paths carved out by massive particles, such as baryons and the dark matter. The fact that light is confined to these trajectories can lead to spectacular visual effects that are not terribly different from the fun house mirrors that Jordan Peele used to great effect in his film *Us*. The way fun house mirrors work is through distortions in the glass that cause uneven reflections of the images in them. A head looks too long or too wide, while the rest of the body becomes minuscule. That kind of weird stuff.

The curvature of spacetime causes it to act like a cosmic fun house mirror. Imagine taking a selfie using the mirror in front of you while there is also a mirror behind you. Then multiple copies of your likeness appear in the picture you take. Spacetime causes the same thing to happen with galaxies. This phenomenon, known as strong gravitational lensing, leads to telescopes capturing pictures of galaxies where there are multiple copies of the same galaxy appearing in the image. You might be wondering how come, if this is a thing, it doesn't happen in our solar system and why wasn't it noticed long before Einstein suggested that people should go looking for the effect.

In fact, there were hints. Scientists like Henry Cavendish pointed out in the late nineteenth century that even Newtonian physics suggested that light should bend around very massive objects like our home star, the sun. But it turned out that using Newtonian physics only accounts for half of the amount of the sun's deflection of light from a straight path—the machinery of general relativity is required for the amount we observe today. Actually, Einstein predicted that there might be some mild gravitational lensing due to the sun, but as yet, we've been unable to test this theory because we haven't sent a machine far enough away. But knowing what we know now, we might expect to see light deflection and not multiple pictures of Jupiter when we look at it.

The key missing piece is dark matter. We observe strong gravitational lensing over long distances when there is a large amount of mass that can act as a lens. For example, galaxy clusters are prime candidates for lensing objects. This lens exists in the foreground, in front of whatever object we are actually imaging, like maybe a very bright galaxy. In this sense, the cluster is like a camera lens that magnifies and distorts how the galaxy looks to us. It does this by bending spacetime so much that the light takes unusual paths to get to us. This can lead to a few different effects: multiple images, the appearance of an apparent ring of light around the galaxy (this is called an Einstein ring), and arcs of light. Gravitational lensing is, effectively, what happens when matter tells spacetime how to move and the movement and shape of spacetime sets out the possible pathways for everything that exists inside of it, including light.

The strong gravitational lensing that I've described involves a galaxy in the background and some massive thing (like a galaxy cluster) bending the spacetime in front of it so extremely that anyone looking at the product can tell that some kind of distortion has occurred. Sometimes gravitational lensing isn't so strong. Something called weak gravitational lensing also occurs. I tend to think of this phenomenon as being a lot like systemic racism. You look at any one incident, say when someone comments on my hair and asks me if it's real, and some person who hasn't experienced racism might say, "Oh, that's not racism. That person was just curious." The hair incident, which happened to me while I was grabbing lunch at an eatery primarily frequented by fellow employees at NASA's Goddard Space Flight Center, is a classic example of an individual manifestation of systemic racism. But in order to understand it as such, one has to have an awareness of systemic racism or a lifetime of experience with its various patterns. If you're experienced, it's easy to identify. I didn't need anyone to tell me that the white man who asked me if my hair was a wig was doing something that Black folks might call "super white" and in academic parlance is a microaggression—an everyday, almost mundane expression of racism.

Noticing the occurrence of weak gravitational lensing requires a similar capacity for pattern detection. Weak gravitational lensing, rather than being really obvious, causes small distortions. The distortions are small enough that looking at any single galaxy might make you think, "Oh, that galaxy is shaped a bit oddly." To figure out that its image is being distorted requires looking at a whole collection of galaxies and checking correlations between their shapes. Is a single gravitational lens weakly causing the galaxies to look longer and thinner than they would without the lens? Another possible effect is something I like to call "bananafication." This is when the galaxy's apparent shape starts to arc a little bit, with a lens causing the image to look bent, a little like a banana.

Observing weak gravitational lensing, like becoming intimately familiar with racism, requires a sophisticated analysis pipeline. It's not something a beginner can do without experience or a plan. We can do statistics over the field and see that there's an effect, and denying the effect's existence, like the phenomenon of weak gravitational lensing, is roughly like denying that physics works. Every incident looks individual to inexperienced outsiders, but there's a statistical pattern that an expert can catch. In the case of weak gravitational lensing, that expert is a person with a good computer algorithm that can do a powerful analysis. Getting this right is so hard that astrophysicists have taken to creating competitions where they give coders manufactured data and tell them to create code that can discern which galaxies are distorted and which ones are not. It's hard. But unless general relativity is extremely wrong (and in the weak regime, it's been pretty well tested), weak gravitational lensing almost certainly happens.

It's easy to argue that identifying racism should be easier than observing gravitational lensing, but I'm not sure the evidence stacks up that way. Even good experts—people who experience racism—sometimes disagree with one another, and people who have a lifetime of experience with racism don't always have a good structural analysis that helps them situate their experience in context. By analogy, someone who knows how to run the weak lensing algorithm may not actually have a good understanding of general relativity. But the biggest problem remains people with no experience or expertise feeling entitled to tell me what is real and what isn't. Throughout my entire life, people who have never experienced racism have argued with me, vigorously, about my tales of racism, questioning my perception and my analytic capabilities. Their inability to see this—and the frequency with which people who have no experience with racism question the observations of people who experience it a lot—tells me that the experts have sophisticated abilities and the non-experts have suppressed their capacity and willingness to develop these abilities. In my view, it is delusional in the face of overwhelming statistical evidence to still insist that microaggressions probably weren't racist.

In the post-2016 election era, one hopes it became increasingly clear just how delusional these claims can be. But even as obvious racist dialogue ramped up, while I was writing this chapter I witnessed a white man tell an Indigenous Canadian scientist that it was rude to call anti–Native Hawaiian comments what they were—racist and colonialist—during a contentious conversation about occupied Indigenous land. In other words, it was rude to say that a debate about the right to engage in colonialism contained colonialist comments. No such pushback was offered to the colonialist commentators. The discursive choices of settler colonials, the very language they speak and their position in any conversation, is always treated as supreme to everyone else's.

I long thought that the primary problem this created was alienating Black people from science and the scientific community. And that *is* a major problem. But I hadn't thought about the other side: like me, other Black people hunger for a connection to scientific thought and will overcome the barriers placed in front of them in order to learn more—but there are real barriers there. In fact, having become an insider to the institutions of science and science communication, I think those who are outsiders are often not able to see the full extent of the barriers placed in front of them. Members of the general public rely on specially trained science journalists and on scientists to translate concepts laden with jargon into accessible language. This is not easy to do. Believe me, I really struggled with that part of writing this book. In the process of translating this language, the audience you have in mind matters. And this isn't just a matter of translating jargon but also developing new jargon. And nowhere has this been clearer to me than in discussions about dark matter.

If the original idea behind comparing Black people to dark matter was specifically to highlight the ways in which Black people and our contributions are erased from cultural discourses, the analogy has also taken flight as a way of identifying with Black as a positive, rather than the negative pallor white supremacy gives it. It is also easy in this context to see why the analogies between dark matter and Black people are popular. For Black people who enjoy celebrating our extraordinary side, and that side is for sure very real, it seems to reify the ways in which we are special and unique. I have long responded to this vicious dehumanization by being a vocal member of the "Say it loud! I'm Black and I'm proud!" crew.

But as Linda Chavers reminded us in her January 2016 *Elle* magazine piece: "Here's My Problem With #BlackGirlMagic: Black

Girls Aren't Magical. We're Human." Chavers's piece caused a stir when it came out; for many Black women, it felt like a punch in the gut. Believing in our own unique power is one of the psychological tools we use to survive a white supremacy that puts us down and diminishes our humanity. Even I, when it came out, told friends that while Chavers had a point, I wasn't sure she should have published it. I see it differently now that I've had time to mature and to step back from the ways I was mistreating myself. In her piece, Chavers notes, "The 'strong, black woman' archetype, which also includes the mourning black woman who suffers in silence, is the idea that we can survive it all, that we can withstand it. That we are, in fact, superhuman. Black girl magic sounds to me like just another way of saying the same thing, and it is smothering and stunting. It is, above all, constricting rather than freeing."

I oscillate between feeling that Black women are extraordinary for putting up with being seen as machines with endless competency, endless power, and most terribly, endless capacity to absorb the worst of humanity, and feeling that it is simply horrible that it happens to us. It's not just something that happens at the hands of white supremacy's most evident beneficiaries and vehicles—white people—but also can even come from our Black women friends. I've had moments where people I knew told me hurtful things and then said they didn't think it would bug me because they knew it had been said to me before. Certainly in those moments, I felt both hypervisible and also like my humanity was being denied because my capacity to feel was out of sight and out of mind. But the hurt I experienced for weeks afterward was a normal human feeling.

The pressure to be infinitely competent is expected of Black women, no matter the cost to us physically, psychologically, or socially. In her essay, Chavers also talks about being a Black woman academic who lives with multiple sclerosis (MS), and the feelings of failure that have come with not being able to compete with people who are not disabled by a world oriented toward people who don't have chronic pain. I don't have MS, but thanks to a car accident when I was 19, I've had a series of problems that kept me confined to bed at various points, kept me from eating as much as I wanted and needed to, and made it difficult for me to keep up physically with other students and practitioners of science. But I always felt pressured to get over this and overcome this, not just because the world tends to disregard the experiences of people with chronic pain, but also because as a Black woman, I had barriers to break. I wanted to be normal, but I felt I had to be extraordinary. In the process of pushing myself, I damaged my body more and at times made my chronic pain worse. To this day, I still struggle with the impulse to work through stupid amounts of pain because I want to be a scientist, just like the white men. Fellow queer theoretical physicist Brian Shuve commented to me that our white, non-disabled, senior colleagues model this behavior and tell us to do this, complete with stories about a certain famed professor and his twenty espressos a day and people not sleeping for days on end just to get an idea out before other people get to it first.

In other words, I am not dark matter, or anything like it. Black people are human. Black people, too, are luminous matter. I don't mean this metaphorically; we are literally luminous. We give off infrared light due to our body heat, just like white people. Light interacts with our skin, making us physically visible to anyone who engages the world through sight—not just police, but also our beloveds. Despite the violence we face, Black people, too, remain curious. Black people, too, make a life in science into their destiny. There is a persistent obsession with demarcating Black people's humanity as substantively different, and usually lesser, than that of white people. It's simply not true though.

I do sometimes feel that I am forced to live my life *pretending* like I am dark matter, like things go through me, like I produce ideas of cultural value and must then watch as those ideas shift things but without anyone ever noticing the source. But at the end of the day, when Black folks invoke the refrain "Black Lives Matter," it is in part a call to see that we are human, just like white people. A few years ago, for example, a couple of white scientists I know and like wrote a review—a summary of everything we know about dark matter. There is an important concept called the "dark sector," which comes from the idea that dark matter might be part of a whole class of invisible particles, rather than a singular anomaly. In their summary, my two friends wrote that they decided to create a hypothetical scientist for the purposes of a thought experiment. This hypothetical scientist would be made of dark sector particles, and so naturally, they thought, the scientist should be called a "dark scientist." As I read this part of the review, alarms went off in my mind. The problem? "Dark scientists" aren't a hypothetical—they're real. People of African, South Asian, Australian Aboriginal, and Pacific Islander descent, among others, come in a lot of shades, and some of them are very dark-skinned. Dark scientists already exist, and many of them are members of my ethno-racial communities. Of course I understood what the writers meant, and I also knew that the conclusion they had come to highlighted a problem with our "dark matter" nomenclature.

When I wrote to my friends about this, they were immediately concerned and quickly updated the paper. I don't bring up this example to make them feel bad. Instead I want to highlight a problem with a phrase like "dark matter." The word "dark" means different things to different people, and in fact, the phrase itself has led to widespread misunderstanding even within the community of physicists and astronomers. Few people ever stop to think that dark matter actually isn't dark. It's just invisible to light. It's transparent—more like a piece of glass than a chalkboard. I think if a Black scientist had come up with the term, "dark matter" probably wouldn't be their phrase of choice, just like I doubt a Black scientist would have propagated the use of "colored physics" in particle theory. On the *PhDivas* podcast—cohosted by one of the very few Black women professors of biomedical engineering, Liz Wayne, and Asian Canadian literary theorist Christine "Xine" Yao—I was asked what a Black scientist might have chosen instead. I proposed "non-luminous ether" because physicists do love a good throwback word play.

Part of what bugs me about the dark matter analogy is that I think the comparison affirms for white scientists that Black people are indeed not really built like they are. This analogy can take things in a bad direction for scientists who understand dark matter scientifically but have no real understanding of how Blackness is constructed or what Black experiences are like. Because of the analogy's emphasis on a material that is completely foreign to humans, it segregates Black people as something outside of so-called normal humanity—which affirms our non-humanness both consciously and unconsciously for people who are unsure about how Black people fit into their worldview, or worse, are certain that we are subhuman.

This isn't merely a theoretical concern. To be a budding Black scientist—a Black science student—is to have people assume you are deficient. I don't just mean that some people will assume that you are intellectually incapable, although there's always that. There will be people like the guy who emailed me and called me a "giddy colored girl pretending to be a scientist." But there will also be the noblesse oblige people, the ones who believe you need to be saved from the deficiencies that racism causes in you. They will not be able to see your strength or what you uniquely bring to the table. Instead, assuming your deficiency is tied to your ethnoracial roots, they focus on trying to teach you how to approximate being white in a world defined by whiteness. My friend Corey Welch, a Northern Cheyenne citizen, biologist, and director of the STEM Scholars Program at Iowa State University, calls this the "deficiency model." This model implicitly suggests that scientific thought is new to Black people or that white people are introducing empirical thought—using sensory information to make rational deductions about the world—to Black people.

This is yet another form of othering. Just as Black people are luminous, normal matter, Black people are also, like the rest of the species, thinkers with a proclivity for the scientific. This is fundamentally what the deficiency model is about: scientific thought is new to Black people, whose intellectual history includes "only being slaves" and then "being primitive villagers" in Africa. I put these things in quotes because even though they are rarely explicitly stated (I can't say they are never stated though), they are implicit in the way that our textbooks teach the history of Western knowledge, as if it is a global history of scientific thought. As I described in "The Biggest Picture There Is," this ignores an extensive, now well-documented history of imperial expeditions collecting Indigenous information from around the world, including Africa, and taking it back to Europe to be collated into the collection of information that we sometimes refer to as "science." Europeans were excellent at collecting information, just as they were excellent at collecting—kidnapping—people. It is up to us not to confuse a combination of collection and invention for invention alone.

Colonialism means that who we really are is invisible specifically because of a lack of trying to see us. Black people of all walks of life are familiar with this experience of both being actively diminished and ignored, and this is another reason that the dark matter analogy troubles me. By drawing on Black folks' own experiences with marginalization, it represents the concept and science behind dark matter to the community in a particular way. Ultimately, this running analogy between Black people and dark matter serves to misrepresent science to Black people who are already often cheated out of access to information about science. If nothing else, it is clear from the powerful way that dark matter has gripped Black imaginations that the word "dark" caught people's attention. The analogy draws on their experiences as Black people to give the phrase meaning, yet this is not the meaning that dark matter should have as a scientific concept. I feel like there is a missed opportunity here to make dark matter our own, but in a way that acknowledges that science is ours too. We have a right to the technical tools that lead us to ask questions like "what is the physics of melanin?" and have the skills to take a deep dive into the literature—or a laboratory—to tease out answers.

Plus, Black people are an audience too. This seems like an obvious statement, but something I often hear is that "we" have a real problem getting Black people interested in science. There's something deeply insidious about the inability to recognize Black people as an audience. The Black community is owed better information and education around science and our historical relationship to it. For example, I think the story of Elmer Imes (which I discussed in Chapter 4) is largely unknown in the Black community. How wonderful would it be for Black children to know his name *and* his science. Importantly, stronger connections with our own legacy in science opens doors for new and better analogies that reflect perspectives on physics that are both hella Black and accurate from a technical standpoint. I don't want to feel caught between Black studies and physics. I want to feel at home in both. I want all Black physicists to have that option. Black scientists can become reference points for the use of physics analogies in discussions about Blackness, rather than, as I have seen, a persistent reliance on the popular science writing of white scientists. These new analogies might also feel less transactional and require less of what W. E. B. Du Bois called a "double consciousness" from Black scientists.

For example, I once wrote an essay on my dissertation topic—as in, my area of expertise—for a leading science magazine. I had gone out of my way to write the essay and speak science in my voice, rather than trying to put on Authoritative Science Voice, which is effectively the same thing as, White Man Confidently Telling You How the World Works. I wanted to share the inherent uncertainty that exists on the boundaries of what scientists understand, which is precisely the place where we do our work. And I wanted to write it as if I was a Black girl who grew up in East LA. The response I received from the editor was that I had taken the wrong tone and the whole thing would need to be re-written. The problem wasn't that it was poorly written—they just didn't think the voice was a "good fit." In other words, being honest about science and especially in my own words was not an acceptable way to communicate to the public about science.

But which public? Let's be real: science writing has rarely been written for Black people. The science writing that does so tends to be about the life sciences, not physics that looks beyond our bodies and to the universe beyond our home planet. Pick up a publication aimed at Black audiences, whether it's owned by white people or not, and you will find no science department. Pick up a publication aimed at those interested in science, almost always owned and run by white people and never by Black people, and you will find few Black writers and little writing that regards Black people as a central audience.

We deserve better. But that requires seeing the world we live in, and how it is structured, for what it truly is. The force that gets in the way—racism—is the weak gravitational lensing of human dynamics. From the point of view of a galaxy or one of its immediate neighbors, if you're not the galaxy you just think it has the shape that it has, not that there's a lensing effect, but the galaxy knows the truth. If you can't see the lensing, maybe you need to adjust your viewpoint. Maybe the galaxy doesn't need to change its size to match your view. No one ever looks at a persistent correlation in science and says, "that's just a coincidence!" But so often scientists refuse to acknowledge persistent correlations around racism in science. Scientists trust each other to speak about statistical analyses of something, but somehow can't take it when Black people talk about patterns of racism, even when our comments are backed up by extensive peer-reviewed research. At the same time, sometimes folks in our own community are shocked by our existence. This can make Black scientists feel erased and some might even use the word "invisible" to describe how it makes them feel. Importantly, we are real and not just the stuff of Afrofuturist imagination, although I hope we are that too.

SEVEN

WHO IS A SCIENTIST?

To be Black in physics is to confront hard questions about whether and where you belong. I didn't meet a Black woman with a PhD in physics until I was almost done with college. I didn't meet a Black woman professor of physics at a major research institution—a role model for what I wanted to be—until I was partway through graduate school and that same woman, Nadya Mason, became a professor (at University of Illinois, Urbana-Champaign). Of course physicists are not the only ones to struggle with the need for role models. In the essay "Saving the Life That Is Your Own," Alice Walker describes her own search: "Mindful that throughout my four years at a prestigious black and then a prestigious white college I had heard not one word about early black women writers, one of my first tasks was simply to determine whether they had existed. After this, I could breathe easier, with more assurance about the profession I myself had chosen."

When Walker looked, she found Black women, and it helped her breathe. I once spent a few minutes with Shirley Ann Jackson, the first Black woman PhD in the field of particle theory and only the second Black woman PhD in physics ever (1973, MIT). By the time I met her though, she had left active research in physics decades earlier, and her career as faculty and permanent research staff was in the field of condensed matter—Nadya's area, and one that is scientifically and culturally very different from particle physics. The year before my PhD was awarded in 2011, observational cosmologist Marcelle Soares-Santos became the first Black woman to earn a PhD in cosmology—not just in Brazil, but as far as I have been able to find, on the planet. I've never met a Black woman professor working in the field of particle physics theory because I am the first. I've also never met a Black woman professor in the field of theoretical cosmology because I am the first—not just the first professor, but also the first Black woman cosmology theory PhD. We are rare and precious, and part of our burden is we must craft our own sense of belonging because we are in spaces where no Black woman scientist has gone before.

To the extent that anyone in my professional community cares about these questions of belonging, of the ability of Black women like me (and unlike me) to live and breathe with ease, people typically respond by talking about things like sexism in the classroom, sexual harassment in the lab, and the people who believe that there are inherent, essential differences between women's intelligence and men's intelligence. In these discussions, we usually accept the essential categories of "men" and "women," although as I discuss in a later chapter, there are many reasons to push back against the idea that the gender binary is at all scientific. Colorism comes up rarely to never. But we also too often ignore the deeper questions at hand: what does it mean to be a scientist? What are the social structures that shape our definitions of "scientist" and "nonscientist"?

To start, I am a science professor, and I don't even know if I am an astrophysicist or not. I am definitely a theoretical physicist, but I also definitely have two degrees in astronomy. Most of the work that I do depends on astronomical observations. This might seem like a strange thing to worry about, but I often get asked if I am an astrophysicist or a theoretical physicist, as if the two are mutually exclusive. I'm a member of both the American Physical Society and the American Astronomical Society, the two professional organizations that represent US-based physicists and astronomers, respectively. This question continues to be a source of confusion not just for the general public but even for people in my family. The truth is that answering this question is not as simple as one might think, and as we go back in time, it gets more complicated too.

The word "astronomy" comes from the Greek *ἀστρονομία*, "astronomia." It is a sufficiently old word that it's not difficult to label people who lived centuries ago as "astronomers." By contrast, the word "physicist" is newer. William Whewell coined it in English in 1836. Before that, Germans had a word, "physiker," that applied to a similar group of thinkers, but I tend to consider physics as not truly unifying into an international discipline with a fully shared set of aims and discourses until the late nineteenth and early twentieth centuries. This leaves those of us who consider ourselves history nerds with a thorny question: is it reasonable to label people who lived before the advent of the word, with the word? Was Isaac Newton a physicist? Even if we take the mid-1700s introduction of physiker in Germany as the starting point of physicist as an identity, Newton had been dead for 30 years by the time physiker arrived on the scene.

Maybe it was Newton's work, which radically changed how European intellectuals viewed the world, that produced the need for such a vocabulary shift. The boundaries of academic disciplines are socially constructed, a product of their times and the people living in them. There was no historical moment where G-d* handed tablets to Moses on the Mount, with a list of academic disciplines and what fell in one or the other. The contemporary categorization of Newton as a physicist is also a social phenomenon: a choice that people make because they believe that were Newton alive now,

---

* As a Jew, I do not spell out the full word when I am referencing Torah or a prayer.

that's how we would categorize him. We will never know, however, whether Newton himself would have approved of physicist as a moniker, rather than a label like natural philosopher. As a deeply religious man—who coincidentally was also a complete asshole—Newton may not have approved of the separation of church and scientific thought that seemed to be unspooling around the time that Whewell introduced "physicist" into the world. While colloquially we are quite comfortable with assigning Newton the mantle of physicist, perhaps contextually it is inappropriate to say he played a role that didn't exist while he was alive.

With European astronomers, however, there shouldn't be so many complications. The word is older than the idea of Europe. If someone did astronomical work, they were effectively an astronomer, right? This is what I had always thought, and I had always assumed everyone agreed with me, until just a few months after I completed the requirements for my PhD and visited my old childhood haunt, the Smithsonian National Air and Space Museum in Washington, D.C. It was October 2010, and I had just moved to the area, having just defended my dissertation. Though my doctorate was not formally awarded until the following spring, a year later than planned because of chronic illness, I had moved to take up a competitive NASA Postdoctoral Program Fellowship in the Observational Cosmology Lab at Goddard Space Flight Center in Greenbelt, Maryland. My Canadian boyfriend had driven down with me from Ontario to help with the move, and he had never been to the D.C. area before. I, on the other hand, had spent several summers there as a kid because my father and stepmother lived just outside the city. The National Air and Space Museum was a childhood favorite, and I was excited to show my partner an important part of my early years.

When we arrived, we learned that the museum had an exhibit on the history of the telescope. Though I'm a theorist, I was super stoked about this. After all, I am a product of the Harvard-Smithsonian Center for Astrophysics, which is one of the homes of the Chandra X-ray Observatory. My undergraduate lab course was in astrophysics and consisted of two projects, one of which required taking observations with the Very Large Array, an array of radio dishes in New Mexico made famous by the film *Contact*, which starred Jodie Foster as a radio astronomer. Though my PhD work had taken me some intellectual distance from observational astrophysics, I had actively chosen a postdoctoral fellowship that brought me into close proximity with it. In fact, my NASA postdoc application had included a proposal to do preparatory work to help a next-generation space telescope do weak gravitational lensing observations that would address the cosmic acceleration problem. Even though I left NASA eleven months later to take up a prestigious fellowship at MIT, what I learned during my time at Goddard has served me well, and I have since become actively involved in proposed space telescope projects.

All of this is to say that I'm not as allergic to telescopes as one might expect from someone who prefers to do pen and paper calculations, rather than data analysis, all day. In fact, I was pretty excited about the telescope exhibit, and I had no intentions of getting into any arguments with any Smithsonian employees about it. I also want to add that I have never been particularly focused on arguing for justice for white women who died hundreds of years ago, since typically they were beneficiaries of the enslaving colonialism that kept my family in bondage. But when I went to the Smithsonian in 2010 and saw in the major exhibit about telescopes that Caroline Herschel was labeled only as an assistant to her brother William, not as an astronomer in her own right, I got angry about it. At the time, I was actually unaware that she had been something of a barrier breaker as one of the first two women inducted into the UK's Royal Society. It didn't matter, though. I have kind of an internal radar for structural sexism—aka patriarchal bullshit—and it went off, fiercely, as I stared incredulously at the exhibit.

I don't remember how I first heard about Caroline Herschel, but it was sometime after college. I don't think it was from Wikipedia, but it may have been. Maybe it was something I learned when the Herschel Space Observatory launched in 2009. I have distinct memories of pointing out to people that it was the William *and* Caroline Herschel Space Telescope, not just named for William Herschel. This didn't seem particularly controversial: while William was recognized as the more famous of the two, it was clear that his sister Caroline trained by his side and on his death took up the mantle of completing his work—or I should say, work he had begun and that she had started to do in collaboration with him, in the process also making some discoveries of her own. I was pleased to have a woman acknowledged as an important historical figure in astronomy, even if it was possible that she wouldn't have seen me as a woman.

Around the same time that I went to the exhibit, I had received my sponsorship to become a full member of the American Astronomical Society and in the process became a member of the Committee on the Status of Women. The committee's newsletter is a collection of links sent in by people who felt they would be of interest to the membership, sometimes with a comment. So I sent a contribution, describing my shock and concern. "Imagine my dismay when I got to the section about Caroline and William Herschel, a sister-brother team of astronomers, and saw their names attached to the following titles: William Herschel: The Complete Astronomer and Caroline Herschel: William's Essential Assistant. The paragraph describing Caroline goes on to begin, 'A fine astronomer in her own right . . .' Well, if she was an astronomer, how come she doesn't get the same label as her brother?" I wrote.

I did not expect what happened next. An editor at the *Smithsonian Magazine* saw my note and decided to respond publicly, not even bothering to contact me for a quote or to discuss the issue with me. In an entry on the magazine's blog, the editor wrote that it was perfectly reasonable to describe Caroline Herschel as an assistant rather than an astronomer in her own right—even though she was in fact discussed as much, at least in the fine print. To add insult to injury, the publication referred to "Chanda Prescod-Weinstein, a local postdoc" rather than "NASA Postdoctoral Program Fellow Dr. Chanda Prescod-Weinstein."* Though my PhD had not been formally awarded, they didn't know that, and they *did* know about the competitive fellowship I had won. While I don't think that titles always matter, in context, the refusal to recognize my expertise or even clarify the nature of my job felt like an attempt to minimize the importance of my stance on how Herschel was being categorized. This was followed by an email from David DeVorkin, one of Air and Space's senior curators, where he justified their choices: "History is about context. If one were to ask Caroline how she would have described herself, I believe she may well have said 'essential assistant' given the gender relations of that day and her personal view of her relationship to her brother."

I was stunned. Having spent half my childhood thinking of the National Air and Space Museum as a kind of home, it felt terrible to return as an expert, only to be dinged by the institution's staff for wanting another woman to be recognized as an expert too. I was even more disappointed to have a woman writer respond so vigorously in defense of what I saw as a demotion. I had imagined that the misrepresentation of Ms. Herschel was an oversight, not one that the curators would feel merited a spirited defense, complete with what felt like a shady comment about my own right to an opinion on the subject. Women who work in astronomy responded in the comments section of the Smithsonian blog by pointing out that assisting is how junior astronomers become independent astronomers. And to this day, reading the responses still irritates me. Caroline may have called herself an assistant, and certainly we should tell young girls and femmes that because of patriarchy she may not have ever imagined herself as anything else. But history does not require us to therefore deny her the label in retrospect. By contrast, we call Isaac Newton one of the greatest

* The text of the online publication was updated only recently, almost a decade later.

physicists of all time, even when the word "physicist" didn't exist until 109 years after his death.

And Caroline Herschel did remarkable things. She was the first woman, jointly with Mary Somerville, to be named an honorary member of the UK's Royal Astronomical Society. She was also the first woman to receive that society's highest honor, the Gold Medal. Later she was made an honorary member of the Royal Irish Academy and awarded the King of Prussia's Gold Medal for Science. I haven't named any discoveries in listing off her accolades, but in some sense I shouldn't need to. A woman being so widely decorated in the sciences was nearly unheard of during her lifetime, 1750–1848. To believe that she wasn't a real scientist and hadn't earned them is to believe that she was just being honored by association, for being an assistant. That's still a ridiculous proposition in 2021 because, as the saying goes, we all work twice as hard for half the recognition, and in 1828, when she received her first gold medal, the men wouldn't have dreamed of it. For millennia, women have assisted men, and it has never been considered worthy of even mentioning really, much less the granting of any statuses traditionally reserved for men.

A decade after my exchange with the Smithsonian, though, I feel compelled to make the case for awards given nearly two hundred years ago to a woman who has also been dead for nearly 200 years. So, for the record, Herschel discovered a comet. All by herself! In 1788. We know for certain that she wasn't just being a hysterical woman imagining shifty things in the sky because a couple of men saw it after she did. It wasn't seen again until 1939 because it has an orbit of one hundred and fifty-five years—meaning that I will likely be gone when it comes back (in 2092).

A few years before the comet, Herschel became the first person to observe a nebula that wasn't listed in any of the catalogues in use at the time. We now know that this object is a small galaxy that is gravitationally bound to Andromeda, the Milky Way's nearest large galaxy companion. NGC 205, as Herschel's discovery is now known, is technically a dwarf elliptical galaxy. Unlike our

Milky Way and Andromeda, it is not a spiral, and instead has an oval shape. Dwarf galaxies, especially ellipticals, are considered to be especially critical objects in contemporary astronomy because they may be some of the earliest large-scale structures to form in the universe. NASA's next generation space telescope, widely known as JWST, will provide crucial new insight into how common they are. In finding NGC 205, Herschel contributed an important building block for twenty-first-century cosmology.

Is that enough to call her an astronomer, in big letters? It should be. Most of the living astronomers I know have never been the first to observe an object in the sky. Tycho Brahe, a widely celebrated sixteenth-century astronomer, is famous for the accuracy of his observations, not the new objects he introduced into catalogues. There has never been any doubt about his historical identity or place in the astronomical canon. (His sister Sophie Brahe, I learned while I was working on this book, was an astronomer who suffered marginalization similar to that of Caroline Herschel.) But I also understand that these debates about her label cannot be disentangled from how we understand her social location in the eighteenth and nineteenth centuries, and they cannot be disentangled from ongoing challenges that women and other people who are not cis men face in being recognized as people with brains.

"You have to be twice as good to get half as far" is advice I think nearly everybody Black in America grows up with. This is probably one reason the misrepresentation of Herschel's record stood out so strongly to me. I had been hearing that and bearing witness to its meaning my entire life. Being confronted with how Herschel was being treated by my contemporaries felt like a warning about how I, a newly minted and formally recognized adult scientist, would be treated. I wondered, "If dead white people aren't even properly celebrated, what chance do living Black women have?"

Of course, the problem of recognition is not specific to science, but rather is endemic to the society in which science unfolds. Increasingly, I find it difficult to imagine addressing the issues that concern me most by focusing so specifically on the context of science. I struggle with the chasm between the fact that being able to think about physics all day is an incredible privilege and the fact that academia is still a capitalist nightmare that takes the life out of people who are conscious of its problems. Everywhere we go, no matter how hard we try to avoid it, capitalism follows us. This might seem irrelevant to Herschel's story, but it's deeply entangled with why I responded so strongly to it.

As Black feminist theorist Imani Perry argues in her powerful history and analysis of patriarchy, *Vexy Thing: On Gender and Liberation*, Black people—and Black women especially—have never been recognized by Euro-American thought as complete humans, much less thinkers who are equal to white people (men) in their competence. Our humanity has always been regarded in fractions, has been established as a subject for legal debate, and has been wholly dismissed in the context of understanding labor and compensation for it. It is not a coincidence that race as a biosocial construct evolved side by side with capitalism. Race was in fact a justification for what South African anti-apartheid activists called "racial capitalism." Cedric Robinson, in his book *Black Marxism: The Making of the Black Radical Tradition*, argues that racialism, the idea that humans can be naturally divided into racial groups, had in fact been the basis for European feudalism and that capitalism is merely an adaptation. In *The Half Has Never Been Told*, historian Ed Baptist built on this argument, noting that capitalism and racial capitalism are one in the same—one cannot understand capitalism without understanding the stolen labor of enslaved Africans and their descendants as a fundamental organizing principle and building block. Black people were never people to capitalism; my ancestors were an energy source. In physicist terms, Black people were akin to a perpetual motion machine: an infinitely cheap source of energy that produced goods and took care of white people's physical needs, up to and including their most basic and violent fantasies, from rape to murder.

I was far into adulthood before I really considered my basic understanding of the psychosocial experience of slavery. Despite being someone who is relatively well educated about slavery, when I thought about what it was like to be an enslaved person, I thought about how harrowing and torturous it was. I thought about how much fear people lived with. I thought about how much physical pain people lived with. As a child I read *Roots* and imagined the pain that Alex Haley's ancestors had lived with, and how my ancestors must have experienced something similar. I also read Frederick Douglass and Sojourner Truth's autobiographies and tried to envision stories of escape. After I spent over a year saving my allowance money, I was one of the first children to own the American Girl Addy doll (which kept me company as I wrote this book), who was envisioned as a girl who escapes slavery with her mother and is for the first time experiencing a proper childhood in freedom.

It was the TV show *Underground* that changed my perspective. When I tried to get friends to watch the show when it was first at risk of being cancelled, I described it as a slavery heist show where enslaved people were steeling themselves for freedom. *Underground* completely changed my view on Black history by portraying enslaved people as thinkers, schemers, planners, and dreamers. In one scene, a character is tasked with building a bridge. It was in that moment that I realized he was an engineer—who didn't know how to read. From that point forward, I understood plantations and other sites of slavery as places of work with not just a skilled workforce, but a workforce with advanced skills. Plantations were made possible by expert, world-class in fact, agriculturalists, engineers, midwives, chefs, and sex workers—who also happened to be enslaved.

This understanding has been critical for me in rethinking how I embrace dual identities of "Black person" and "scientist." For much of my childhood I heard about Black scientists as a people of the future. I didn't hear about Benjamin Banneker until I was in middle school, while I was visiting Washington, D.C., the city he helped plan using astronomical observations. When I applied to Harvard College, I remember explicitly writing in my application that I knew I would be one of the first Black American women to earn a degree in astrophysics. I didn't realize that (according to NSF statistics) I would eventually be the only one in the country to earn a bachelor's in that subject in 2003, but I knew the numbers were low. I had learned to think of science in terms of formal recognition—histories that were written by white people, degrees that were granted by institutions that white people acknowledged. As my understanding of the world of science beyond Europe and its settler colonial satellites evolved, I started to understand that this perspective didn't make sense. But when I thought about science outside of the professionalized, establishment context, I focused on peoples in the continent of Africa, in Asia, and the Pacific Islands, and peoples Indigenous to the Americas. What changed with *Underground* is that I began to synthesize a new, more complete understanding of how the Black Atlantic fits into this picture as kidnapped Indigenous Africans. My whole life I had assumed that there were no scientists in my family history, but then I realized that in fact I had no idea.

Around the same time that I was thinking these ideas through, the film *Hidden Figures* came out. That film portrays a story that I had become familiar with a few years before thanks to a post on Nichelle Gainer's *Vintage Black Glamour* Tumblr about Melba Roy, a Black woman mathematician who worked as a mathematical space scientist at NASA. Roy was one of several Black women whose histories were completely obscured by NASA. My friend Duchess Harris, granddaughter of a hidden figure and a professor at Macalester College who was part of the research team that uncovered this hidden history, tells me that initially even NASA didn't believe the women had existed until she showed them a mid-1960s letter on NASA letterhead that thanked her grandmother for twenty years of service, implying she began in the 1940s. I remember talking to Nichelle about the post and noting how shocked I was that as a Black student in astronomy, I had never heard of Roy. I was even more surprised later when I found out there were more women like her.

It begs the question: what other history is hidden from us? My friend LaKisha Simmons writes in the introduction to her book *Crescent City Girls: The Lives of Young Black Women in Segregated New Orleans* about the difficulty of writing about her subject because Black girls' lives were often not considered worth archiving. Historians reading this will feel like I am stating the obvious, but because I trained professionally as a physicist, I was late to an understanding of why doing Black history is so difficult. Racist kidnappers and enslavers only recorded information about Black people that was relevant to their value as capital—names, ages, assigned gender/sex, any special skill set with specific commercial value. In other words, Black history in the Americas is full of Black women who are hidden figures. Benjamin Banneker had no Black women contemporaries because Black women contended with the double bind of anti-Blackness *and* patriarchy. But there are many reasons to believe there were Black women who thought about astronomy. We know from historical records that enslaved people—potentially including Harriet Tubman—used the North Star for their escapes to freedom. But also, more fundamentally, if I could be curious about the way the sky works, then surely others have had those same questions, even if they didn't have the same conditions that would allow them to fully explore and experience that curiosity.

There are some stories we know about enslaved Black men who made significant contributions to science. One instance that is particularly memorable for me as an alum of Harvard College's Mather House is that of Onesimus, who was purchased in 1706 by Cotton Mather's congregation to serve their minister. Mather, who was a vicious racist, didn't trust Onesimus but came to grudgingly accept that he was intelligent after Onesimus explained how to prevent smallpox. In other words, Onesimus introduced Mather and his New England community to vaccination. As I write this book, I wonder why Mather House continues to be named after a slave owner, Increase Mather (Cotton's father), and not the survivor of slavery who was also a scientist that saved lives. Onesimus House has a great ring to it, and though I've sworn off Harvard reunions, I'd go back for a renaming.

Caroline Herschel was born just a few decades after Cotton Mather died, and it's necessary to contextualize the life she led with how she was able to lead it. Her father made his living as a military musician in what we now call Germany but was at the time part of the Holy Roman Empire. Caroline herself trained as a singer while her brother William was an organist. They migrated to England, where William worked to reshape himself from musician to intellectual and natural philosopher. I mention this because they were beneficiaries of the feudal system that preceded capitalism, but they were also among the growing English middle class that would have benefited from the colonialism and enslavement taking place in the Americas. Who knows—maybe Caroline or William ate sugar made from cane that was picked by one of my ancestors in Barbados. Certainly William's income as a musician, which came in part from landed gentry, was likely tied up in both feudalism and slavery.

Importantly, the same forces that work to deny Caroline Herschel her status as an astronomer worked to extract resources and labor from European peasants and kept my ancestors in bondage while justifying the displacement of Indigenous people from their homes in the Americas. This is a connection that white women who follow in Caroline's footsteps are socialized not to make, although it is something that a well-known contemporary of Herschel, Jane Austen, tried to make clear in her own writing, particularly the novel *Mansfield Park*. When I tell people about this interpretation of Austen's longest and most difficult novel, they are often surprised. This is partly because most readers are not familiar with The Supreme Court of England's 1772 *Somerset v. Stewart*, in which Chief Justice Lord Mansfield ruled that slavery was not encoded in English common law and that therefore enslaved people who were brought to England and Wales could not be held captive. Today, we can read Austen's *Mansfield Park* as an explicit reference to this famous case that was decided just three years before her birth. The novel, in my view, is a rumination on the landed gentry's attempts to maintain their financial status through the ownership of plantations in the West Indies—the Caribbean. With her main character, the poor cousin Fanny, Austen specifically juxtaposes poor white women's experiences with servitude in a genteel context against the uglier but distant captivity that made the lives of all her characters, even the poor ones, possible.

It is this reading of *Mansfield Park* that allows me to be both sympathetic to Caroline Herschel's social location and at the same time careful not to valorize her as a hero without context. Herschel, for all of the ways in which I understand her marginalization, was still a white woman in a time period when being white meant something specific. We are still living in that time period, and as I examine my back and forth with the folks from the Smithsonian Institution a decade ago, I am struck by how much my ways of being in community with white women in science have shifted. I am increasingly aware that as much as I have tried to protect Caroline Herschel's legacy, the likelihood that she would have even thought of me as a human being is possibly low. I once worked hard to identify with white women as fellow women/gender minorities in science, but repeated experiences with racism at the hands of white women in astronomy and physics changed that.

This was exemplified during the height of the 2015 discussions about the Thirty Meter Telescope (TMT) on Maunakea, when kanaka maoli protectors were opposing the building of a large telescope that would be the fourteenth on land that has historically been and continues to be considered sacred. I remember noticing a stark difference between the things white women said about it and the things others, including me, had to say. Repeatedly I witnessed women of color, particularly Indigenous women, questioning the astronomy community's orthodox stance that the building of the TMT was an inherent social and scientific good. I also saw them question the value of building the TMT when it was clear a vocal community of kānaka maoli were opposed to it. We struggled with fears of betrayal—of our values regarding solidarity with fellow oppressed peoples. By contrast, I found that white women struggled also with a sense of betrayal, but they seemed more concerned with betraying the astronomy community than betraying their cultural values. Why? Because their home cultural values often were the same as those of the astronomy community. In one distinct example, I remember a white woman who is involved with a program for underrepresented minorities commenting that the Native Hawaiian protectors just didn't know what was good for them. She hasn't said a word to me in the five years since I told her this was a racist comment.

My point here is not to demonize the figure of the white woman in astronomy, but rather to complicate the idea that because white women are victims of patriarchy that they are not also sometimes perpetrators of white supremacy, with implications for people of color in the sciences. Some of my fiercest allies during that 2015 fight over Maunakea, besides the protectors I got to know who also became very protective and supportive of me, were white women. For example, my deep friendship and political collaboration with astronomer, telescope builder, and white Jewish activist Sarah Tuttle began in those days. Our relationship evolved this way partly because time and again, Sarah has been willing to take up difficult conversations by my side, rather than leave me to manage them alone. I think she and I would continue to be defensive of each other's right to exist, with dignity, even if our friendship somehow became untenable. In this context, I am able to understand my defense of Caroline Herschel as necessary solidarity. Much as I think she would not have defended me, I believe it is important to hold fast to her place in history as an astronomer. The erasure of her role in history is a threat to all of us, and we must hold the line against it.

PHASE 3

THE TROUBLE WITH PHYSICISTS

There's the universe and then there's the process of developing a mathematical understanding of it. This is physics, and physics is a human process.

Physics is not the universe. Rather, it is one very human attempt to get at its innards.

EIGHT

LET ASTRO/PHYSICS BE THE DREAM IT USED TO BE

> My response to racism is anger. I have lived with that anger, ignoring it, feeding upon it, learning to use it before it laid my visions to waste, for most of my life. Once I did it in silence, afraid of the weight. My fear of anger taught me nothing. Your fear of that anger will teach you nothing also.
>
> —Audre Lorde, "The Uses of Anger: Women Responding to Racism"

I WAS ONCE A BLACK CHILD DREAMING OF A VERY SPECIFIC tree of knowledge: particle physics. At the age of ten and a half, I fell in love, thanks to Errol Morris's documentary *A Brief History of Time*. I had not wanted to go see the film because I thought documentaries were boring stuff for old people. Yet my mother had driven us from East LA to the Westside just so we could see a matinee at Laemmle Theatres. It was the story of a man who spent most of his life in a wheelchair and talked using a computer that he controlled with a finger he could barely move. This man, Stephen Hawking, was a university professor at Cambridge in the UK, and he used math to understand how things in the universe worked. The film positioned Hawking as an inheritor of the mantle of

Einstein and though I knew little about Einstein's actual work, I knew he had been an important scientist, one who had changed the way we see the world. Hawking's most important contributions ultimately were to our understanding of black holes as quantum mechanical objects, and partway through its runtime, the film lands on a question: what happens at the center of a black hole? It turned out Einstein had been unable to figure it out, and Hawking was on a mission to develop an explanation.

I thought: *Wait, you can get paid to try and figure out things that Einstein got stuck on?* It was an "aha" moment for me. I too wanted to know what happened at the center of a black hole, and I could make a career out of trying to find out. As we left the theater, I begged my mom to buy me a copy of Hawking's book. Although she had brought me to see the documentary specifically because of my expressed interest in my science elective—as well as my tendency to want to know how to do every sort of math calculation that I learned about—she refused. She told me it might be too hard to read. Her older brother, my Uncle Peter, defied her and got me the book for my eleventh birthday. It took me years to finish, but it remains one of the most memorable and important gifts I've ever received.

A few years later, when I was in high school in West LA and had a three-hour roundtrip school bus commute, and, just as I described in "I ♥ Quarks," I used to regale people with tales of the particles I read about in Stephen Hawking's *A Brief History of Time*. I explained what a quark was, even though I had no idea what it was. I explained the lepton family, even though I didn't have a clue about it. I was very excited about my future as a Harvard-educated particle physicist and cosmologist. I had it all mapped out, thanks in part to a Cambridge University graduate student of Hawking's, who responded when, at age eleven, I had emailed Hawking and asked how to become a theoretical physicist. Go to a first-class university for your undergraduate degree, he wrote me. Then attend a top PhD program. Afterward, I should become a professor.

Then I too would be like Hawking. I researched the best schools for physics in the United States and came to the conclusion that I ought to attend Harvard or Caltech for college. In April of 1999, Harvard and Caltech were two of only three schools that I could afford, despite being admitted to 12. I chose Harvard because it was in a city, it was far from home, and the Harvard-Smithsonian Center for Astrophysics was home to hundreds of astrophysicists and located in a charming part of Cambridge a little way away from the pomp and circumstance of Harvard Yard.

I left East LA just a few weeks after I turned seventeen because someone like me was free to leave. I was a good test taker, a voracious student who enjoyed school, and a child who, because of a combination of luck and colorist colonial social structures, hadn't gotten caught up in the mass incarceration dragnet. When I left for college, I was not completely naive about the challenges I would face as a student of physics and astronomy, what I sometimes call astro/physics for short. I knew that the chips were stacked against women. I expected I might stand out because as a light-skinned Black woman who was often mistaken for a Mexican American when I was close to home, I looked different from the white women who dominated women in science spaces. But I believed the multicultural diet early millennials had been steadily fed that the world was better now and barriers were actually *begging* us to break them.

I didn't understand that I would stand out because I came from a different world, one of cholos from Metro 13 and Bloods and Crips but also Black and Brown girls with lips boldly outlined in thick brown lipliner and pre-crossover Shakira blaring loudly as folks drove down streets with brightly colored houses that were near train tracks and cancer-causing factories and billboards that hadn't seen a word of English on them in years. I didn't understand that I would stand out in my professional community for years on end, even today, because I refuse to decide that I am no longer of East Los, even if it has been half my life now since I lived there. I did not know how lonely I would feel, and the grief I would experience as my ability to speak Spanish has dissipated, that my only opportunities to speak it would come with, as I moved up in my professional world, increasingly frequent visits to four- and five-star hotels staffed at the lowest-paid levels by people who live in neighborhoods like the East Los I grew up in: El Sereno, Lincoln Heights, and Eagle Rock.

At Harvard, I learned that to look professional I would need to wear J. Crew, but it would be many years before I could afford it. After Harvard, I learned to wear big hoop earrings not just because that's how we do but because that's how I consciously put faculty hiring committees on notice about my political commitment to Black femmes and to Black womanhood. I learned also to name the academy as a place where the ruling class is produced, sometimes through psychological warfare, even as it is also a place where we learn wonderful things like the fact that in the Orion Nebula, water naturally produces masers, which are the radio version of lasers.

People I went to college with now tell me that East LA is becoming cool and the *Los Angeles Times* now celebrates property values in the same place that I was told in college was "the bad side of town." East LA certainly wasn't cool the day I got into Harvard at age sixteen, when two white boys in my class, one of whom cheated his way to a high ranking in the class, sat me down and said no matter where I went to college, they would always be better than me. They were only the first and not the last people to insinuate to me that no matter what I did, who I was meant that my humanity would never measure up to theirs. I was probably in the bottom quartile of my high school class socioeconomically, and because of this my dream of becoming a scientist was both a revenge fantasy and a triumph fantasy: I wanted to come home a hero, the local girl who went to Harvard and came back with the magic of particle physics in her hands.

Dreams get deferred. In January 2019, when I became the first Black woman in history to hold a faculty position in either theoretical cosmology or particle theory, I couldn't have been geographically and culturally further away from the grand homecoming that I dreamed of. Almost everyone who gets a PhD in physics must eventually take a job outside of academia. There are simply not enough positions for everyone. Getting a faculty position at all is a big deal, even for white men. Of course Black women (and Black men) are not hired into the ranks of faculty at a rate that is proportional to our numbers in the population, which means that for someone like me, breaking into one of these roles is a steep challenge. It turns out that for various reasons, it's hard for someone like me to get a job at a physics PhD–granting university in Los Angeles—or any highly ranked university in the country.

When I say "someone like me," I specifically mean a working-class Black queer femme from East LA with a politics that remains committed to those roots. That's just another way of saying that I have a big mouth, and I'm not afraid to use it to critique white supremacist heterocisnormative ableist patriarchy. It's also another way of saying that I found that college science programs, PhD programs, postdoc hiring committees, and faculty hiring committees expect candidates to be shaped like a certain kind of white male square, and I am definitely a different geometry. There are Black folks who get by in academia by keeping their heads down, although the extremely low numbers of Black women in physics indicate that keeping your head down isn't enough to get by and succeed. But speaking up is part of what makes my life in physics livable.

What's supposed to keep you going is your love of science and your love of the process of learning science. And while I have been able to hold fast to an almost irrational, near primordial love for cosmology, the astro/physics community has rarely let me enjoy the science. It began with an undergraduate curriculum that was designed for more highly resourced high school graduates than me, the adviser who spent four whole years worrying in front of me that I wasn't smart enough to be a theoretical physicist, and a school full of people intent on reminding me that I didn't look like a physics major (for a bunch of smart kids, what an uncreative response to that piece of information).

My first semester of college felt like drowning. I performed so badly that I called my mom and said I was switching to anthropology. Like a highly recognizable Black West Indian American mother trope she said, "I did not work a second job so you could stay in that nice high school just to announce you are quitting on your dream." My mother will tell you that I am a lot like her, which means that I often don't listen to what I am told, even by my mother, but I heard that message loud and clear: I owed it to the ancestors to try. And because I am a lot like her, I am stubborn in the face of people who tell me no. For four years, I told myself it would get better later. At the same time, my undergraduate adviser was telling me that I should shift my plans because someone with such low grades would never make it as a theoretical physicist. Having no one to counter the noise I was hearing, I partly believed what I was hearing, that I wasn't smart enough.

Through all of this, I was supposed to be enjoying the wonders of physics! Because physics is super fun when your study group is treating you like a stupid charity case because they never bother to remember that while they were enjoying leisurely study time in the library, you were at your work-study job. Never mind that this was a job that the richest university in the world could easily have avoided making it financially necessary for me to take, but it instead loved to remind me was proof of my investment in my education, as if busting my ass and getting in without the privilege of a $2.5 million donation to buy my way in wasn't evidence enough.

Physics is actually not super fun when you start to believe the stories that people are telling you about yourself, just a little, and you hate yourself for being that thing they keep telling you that you are; when you're Black, queer, from a working-class background and your classmates and eventual colleagues regularly say racially insensitive and blatantly racist, homophobic, and classist things with an occasional sprinkling of sexism and misogyny for good measure. When I moved on to grad school, I learned that sometimes white male postdocs are racists and rarely does anyone with influence or power really want to do anything about how those people shape the lives of students like me. That doesn't change much when you're a postdoc, either. In the end, it's not fun to go to work or to visit a department and brace yourself for the regular barrage of questions about your race/ethnicity/Jewish parentage because apparently your skin color screams, "different but ambiguous—stop scientific discussion to ask invasive questions and categorize!" to various white faculty who have power over your career. And of course there's also the part where, if you get through all of those hoops and succeed as an academic, you don't really get to choose where you live, and if you're a Black (Caribbean American) Jewish queer woman, finding community will be incredibly difficult in most small college towns. You're not supposed to mind being the only one, not just on campus, but in the geographic region.

I'm not going to catalog every terrible thing that happened to me as a student or as a postdoc even though I haven't actually mentioned the worst stuff, but the point is that I didn't and still don't fit into the dominant astro/physics culture, and this has really sucked a lot of the fun out of physics for me. White astro/physicists in particular need to understand and take to heart that this is in fact a real issue that doesn't magically go away with admissions and diversity initiatives that fail to address underlying cultural, *structural* issues. I genuinely believe my adviser is both sorry about the advice she gave me and also that in her own way, she was trying to protect me from the brutality of a particle theory community that I still idolized and that she knew was deeply racist and sexist. And her impulse was right—people like me are not particularly welcome in particle theory. Particle and cosmology theorists tend to place a higher premium than other physics researchers on grade point average, access to advanced high-level classes, and one's score on the GREs, the graduate school equivalent of the SAT. Someone like me—someone from communities that have been disadvantaged, who didn't have the audacity to believe that admission to Harvard meant she was one of The Chosen, and who was actively critical of the ruling class and felt it was important to spend time challenging it—was not meant to be a theoretical particle physicist.

At this stage, you might ask, "well, why are you still here?" As I have gotten older, it has become harder to answer that question. There have been times when I felt that the astro/physics community was actively ruining the stars for me because of its unwillingness to detach from dehumanizing ways of making science happen. I can work on trying to enjoy myself more, I can work on shutting out the noise more, I can try to hold on for dear life to the occasional person who gets it, but ultimately the whole community urgently needs to start working with me and for people like me. And with far too few (but highly appreciated) exceptions, it is not. Important bonds that have helped hold me together include my 'Nati Crew, formed through Black scientist community connections and most importantly, the #BlackandSTEM Twitter community founded by biochemist and biophysicist Dr. Stephani Page. I also am lucky to have a spouse who has taught himself to be the coach I need for moments of doubt and panic. Even now that I have a faculty position, I still have many of those. They are just as frequent as before, but they have certainly taken on a different tenor now that I have students depending on me in ways that I didn't when I was "just" a postdoctoral fellow.

For many years, I didn't know if I would succeed at obtaining a faculty position that had the prospect of tenure attached to it, and there was certainly no shortage of senior people who told me I never could. In fact, the last person who really went out of their way to tell me that I couldn't make it as a scientist was one of my most important mentors and professional references—and sadly, also a member of a marginalized group. I don't particularly have the impression that white men get told this a lot, but every Black woman physics PhD I've discussed this with had someone in a position of authority and influence tell them that they weren't cut out to be a physicist. Or they might say, "Your success is proof that the system isn't rigged," when of course one token, or even a few, getting their foot through the door doesn't signal that the playing field is level. Besides, I still don't have tenure. My career and the success I have had in it isn't permanent until then. And the way things are going now in higher education, tenure might not even hold the job security it used to.

Will I stick around in the long run? I don't know. When I go up for tenure, I will find out. Because I was struggling to enjoy the work, because of the very real discrimination I faced, and because of random bad luck that can happen to anyone, I struggled in my early years to be as productive as my peers, or rather to learn the productivity habits that ensure success in my field. Does that signal that I don't have the capacity to make significant contributions to the field—and haven't already through my vocal advocacy for people and dark matter, on top of my carefully thought through scientific publications? Not particularly. Will it be read that way? It often has been.

Despite everything, I have been able to make technical contributions to the field. In fact, my work has recently been recognized by the American Physical Society with the 2021 Edward A. Bouchet Award, the citation for which reads, "For contributions to theoretical cosmology and particle physics, ranging from axion physics to models of inflation to alternative models of dark energy, for tireless efforts in increasing inclusivity in physics, and for cocreating the Particles for Justice movement." I believe that signals that I could contribute even more if given a more equitable chance. My Black, Indigenous, and Brown Latinx colleagues and students and the future aspiring astrophysicists that I haven't met or who haven't been born yet could do really wonderful things, if given a more equitable chance.

While I know I have played a pivotal role in shifting conversations about equity in physics, this work is both seriously devalued in our community and often comes at a personal cost. Scientists don't like being told they're wrong, and white people don't like being told they're racist, so you can imagine how white scientists take hearing that they are contributing to dangerous social structures. This will all be a factor in my tenure case, when my department will select members of the community to share their thoughts about me in confidential letters that discuss my impact and reputation.

In addition, we should critique the types of productivity that are rewarded in theoretical physics, where scientists are encouraged to think superficially and publish every little idea, rather than to think deeply, only publishing things that they can genuinely get behind. Quantum gravity theorist Lee Smolin effectively wrote a whole book about problematic rewards systems in physics called *The Trouble With Physics: The Rise of String Theory, The Fall of a Science, and What Comes Next.** Speaking to the *Guardian* (UK) in 2013, Nobel Laureate Peter Higgs echoed Smolin's concerns about the direction things have gone in, saying, "It's difficult to imagine how I would ever have enough peace and quiet in the present sort of climate to do what I did in 1964." Physicists celebrated the triumphant affirmation of his Higgs particle theory when the particle was detected by the Large Hadron Collider in 2012, but the community as a whole ignored his message about the conditions needed to produce the theory in the first place.

They want his physics, not his thoughts on how physics ought to be done. And if that's how they feel about what a white male Nobel Laureate has to say, they likely could care less about my views, my love notes to science. Quietly people will agree when I say our hurried publish-or-perish culture is damaging to science,

---

* This book inspired my first work of public writing about diversity and inclusion in physics, which fellow theoretical physicist and science communicator Sean Carroll published on his blog back in 2006. In that essay, I point out that for all of the talk about the stifling of diverse ideas in physics, there wasn't enough discussion about the stifling of marginalized people.

but I've never seen anyone in a position of power actually do anything about it. Our academic and economic structures are set up with capitalist incentives to keep it to yourself when you realize something is wrong and to favor quick, superficial work over work that requires deep, plodding thought. Those who get a little power within these structures are rewarded for their silence.

Alongside my struggles to build a career in physics, I regularly grapple with existential challenges. I don't know how to deal with a community that won't acknowledge its colonialist and white supremacist past and present, which virtually guarantees—nay, insists on—a colonialist, white supremacist future. I don't know how to deal with a community full of people who can hear a Lakota woman and a Black woman say that the community discourse about an experiment like the Thirty Meter Telescope on Maunakea is hurting them and not have a large number of people stop and say, "Wait, let's rethink what we're doing here, or at least consider how we're talking about it and what the words we are using mean to different members of our community." I don't know how to deal with a community where Indigenous students can bravely challenge a racist email, only to be told by their classmates that their anti-racist stance was embarrassing to the community.

In our community it is always better to talk about "diversity" and "inclusion" than to use clearer language like "ending anti-Blackness." In *Race on the Brain: What Implicit Bias Gets Wrong About the Struggle for Racial Justice*, legal scholar Jonathan Kahn provides a useful overview of forty years of literature on affirmative action, racism, diversity, and implicit bias. Kahn makes a compelling argument that, fueled by scientism in the cognitive psychology and legal communities, implicit bias has become a master narrative of American race relations that displaces a focus on the unique history of American white supremacy in favor of a focus on the value of "diversity."

What I came to understand while reading Kahn's book is that when we separate legal and social thought about race from

America's history of enslaving and killing people in large numbers, the only timeframe in conversations about race that matters is now—as if it is not contextualized by everything that came before. Rather than considering how a history of structural racism totally and completely shapes the trajectory of a Black child's life, there are those who choose to speak only of whether any given particular microaggression is in fact actually an instance of racism. In this context, assessing experiences of bias requires only considering the intent of someone who might be biased. When the history of forced servitude and genocide do matter, then all of the time from the beginning of those historical phenomena until now matters, and in assessing experiences of bias, the impact on history's victims of bias is salient.

Science has instead wholeheartedly embraced the view that intent matters more than historical context or impact. This is completely dehumanizing because it means victims' feelings always come second. I often don't know how to come to terms with a community that cares little for the full humanity of those working in it. I don't know how to thrive in a community full of people who don't understand that what they are demanding of us is that we assimilate to their sensibility of what's "fair" in science. Perhaps worse are people who do understand that this is a demand for assimilation and think that this demand is ethical. It is destructive and dehumanizing to work in a community that values diversity for the "sake of science," but doesn't value equal opportunity for the sake of respecting other people's fundamental humanity. The implication is that it would be fine to treat people like me badly if we aren't providing valuable intellectual capital.

In this context it's not surprising that many of us experience physics, as a social environment, to be fundamentally empathy-deficient, dominated by people who are generally disinterested in the racism, sexism, homophobia, and classism that I deal with and that impacts my life twenty-four hours a day, not magically only during the hours when I am not working. Our training teaches us to be this way—we're not encouraged to learn other modes of thought and being. On top of that, everything is organized around the idea that my body will operate like some kind of ideal machine. In this profession, my struggles with chronic pain are treated as an inconvenience rather than part of a spectrum of physical and psychological diversity that requires elasticity in our systems. We live and work in an environment that is rigidly committed to an exclusive and sometimes physically violent status quo. How do I make peace with a professional space where one of the most famous departments in the world was also home to three known sexual harassers, and was like that for decades before press articles that preceded the Harvey Weinstein story forced their hand? Faculty insisted they didn't know or couldn't do anything. No one takes responsibility.

I understand intellectually why my work on inflation and axions is interesting, but I am tired of the disjointed feelings of liking the ideas but finding it hard to breathe in the community in which I have to share them. Because I am human and not an objectivity machine, those two things are both part of doing science. Astro/physics loses so many people, including white men, because the community refuses to acknowledge this dual reality, that science is both a mathematical process and a human, social enterprise. Many of my colleagues refuse to consider what it's like for someone like me to watch as the very few Black people ahead of her in the pipeline disappear from the mix. We lose people because too often there aren't enough victories and moments of joy and fun for those of us who are from communities that history kicked to the cotton fields, trails of tears, unsafe street corners, prisons, and deaths at the hands of state actors and vigilantes.

It can be hard to see the wonders of the universe through the social crud. Yet, here I am. As unattractive as the professional world of physicists can be, I am still completely dazzled by quantum field theory. Every year I feel like I am learning particle physics for the first time, and it's still amazing. Spacetime is still curved.

I'm still more fascinated by that fact than I am horrified by all of the terrible behaviors I have witnessed from people who are experts on that fact. And I've had to think about what it means to stay in this particular social configuration of spacetime. I—and so many others—still want a chance to regularly have uninterrupted fun doing science. That means everyone with the power to enact change—everyone with high social capital and relative job security—has to do the work of making this field different. Making it so that I am not constantly putting out fires on all fronts, quietly soothing the concerns of students who are afraid to speak out publicly while pushing back publicly against some of the harmful crap that they are hearing while also trying to let those students know that it's not them, it's the hand they were dealt by a community hell-bent on avoiding, dodging, and stopping change.

Just because I'm good at all of those things doesn't mean this stuff doesn't sting. It more than stings. It hurts. Some of the time I could be spending working is spent protecting my community. Lots and lots and lots of people will respond to this by saying that I need to be more selfish. But do people imagine what that decision is like? To actively say to myself, "I could help people like me and protect them from some of the bad experiences that I wasn't protected from, but I'm not going to do it!" I want to be able to like myself when I go to sleep at night, and selfishly ignoring the suffering of other marginalized people is not part of my personal vision of integrity. At the same time, I increasingly have to make hard choices because I am only one person, and I can't be responsible for every problem. It's not physically possible.

I still like math and the potential it holds to help us craft a compelling cosmological tale. I still think the times table is a miraculous thing, thirty years after I first learned it. I still love that we can use math to understand and describe the history of the universe itself. And I want little children of every shade, gender identity, sex identity, ethnicity, sexual orientation, romantic orientation, (dis)ability, and religion to have access to that cosmos, to have fun with it, to find joy in it. And I still want that for myself. Do our lives and our ability to breathe, not just physically but also intellectually and emotionally, matter? I believe that the cosmos don't belong to straight, white, abled cis men alone, even though for the last few hundred years the scientific community has been organized around uniquely entitling them to knowledge of the cosmos, and even though for much of that time, that knowledge has been used against people like me. I share Carl Sagan's humanist belief that one of our greatest strengths as a species is our transcultural impulse to weave stories about our origins. Imagine what we marginalized people could do if we were no longer marginalized, and studied and worked within a community of support. Imagine.

Here I am, according to Dr. Jami Valentine Miller's list of African American Women in Physics, the fiftieth-something Black American woman to ever earn a PhD from a department of physics, a theoretical particle physicist and cosmologist who graduated from college with what could be called a B minus average if we are being generous, and I still feel shame about this. It feels strange to recall how all of this has made me feel—and to recall all of my feelings about what an awful and often abusive place academia is—while people are locked up in freezing cages in Brooklyn and Laredo and while children are being disappeared by the Department of Homeland Security and the bad part of the US Department of Health and Human Services (which has other offices doing important lifesaving work). In many ways, I lead the life of a highly successful token. I make way more money than my mother raised me on, and I am recognized in the press as a Black woman who is changing the face of science. Black women are outnumbered in physics by Black men at a ratio of 2:1, and Black people overall are outnumbered by white people at a rate that I couldn't bring myself to calculate. My presence in the field represents a broken barrier, and I often feel that I broke that barrier with my bare hands, even though I know the ancestors began the work long before I was born.

Those African ancestors were forcibly separated from their parents and siblings during, before, and after the Middle Passage. Families were torn apart even in the search for freedom. Sojourner Truth had to leave most of her children behind when she walked away to freedom. We talk about Harriet Tubman using knowledge of astronomy by following the drinking gourd—the group of stars scientists call the Big Dipper—to make dangerous trips below the Mason-Dixon Line to lead people to freedom, and perhaps we don't emphasize enough that more than one of those trips was to liberate members of her family. Today, children and their parents march miles and miles under the same stars that Harriet Tubman may have used to navigate, and they too are looking for sanctuary and freedom from violence. While the violence that these Black, Indigenous, and white Central American and Caribbean refugees face has different contours than the violence of chattel slavery, the fact of running for your life by any means necessary, including by foot, is something they have in common. We cannot talk about the wonders of the night sky without talking about the fact that people are running for their lives beneath the same celestial structures that I get paid to think about every day. I do not want to wait to find out how this story ends if we don't get in the way because I know, and as a Black Jew feel in every fiber of my being, what happened when Germans did the same.

My personal success will not end the structural racism that keeps so many Black people and refugees, especially single mothers, their children, and trans folks, in poverty. I also don't believe there is a college or university or any workplace that is a healthy intellectual environment for Black women, and that will remain true as long as capitalism structures our social relationships. I believe that can change, but as my friend Kiese Laymon likes to say, people have to be ready to do The Work to make it happen. The Work requires not just taking action in our professional spaces, but in the wider world. As long as racist poverty remains a fact of our society, academia, including academic physics and astronomy, will remain the mess that it is.

But there are other possibilities, like dreaming Black scientist freedom dreams. Too often, arguments for expanding access to science are made in the name of exceptional people among those being brutalized. Scientists and policy makers focus on the question, "What if that refugee child could solve dark matter if we just gave them a chance?" This framing is ultimately about what value the refugee child has to American intellectual economies, too close to the logic of slavery for my comfort. It is essential that we reject this framework and instead put our energies into eliminating the conditions that cause people to flee for their lives. At the same time, we must question our collective investment in the idea of borders and deny—completely—the value of walls, like the ones both here in the US and Palestine.

My freedom dreams are big. Caring for humanity means ensuring we all have food, water, shelter, health care, equal treatment under the law, and freedom from violence. Sustaining what makes us human also means protecting our deeply human connection to the universe and our impulse to tell stories about it. For we are a hybrid, storytelling species, Sylvia Wynter's *homo narrans*. Access to a dark night sky—to see and be inspired by the universe as it really is—should be available to everyone, not a luxury for the chosen few. We must demand liberation for all, including the right to know and understand the night sky, not as the context of desperate and dangerous searches for freedom, but as the beautiful place that holds the answers to how we came to exist at all.

NINE

THE ANTI-PATRIARCHY AGENDER

Particles themselves are nonbinary.

—Amrou Al-Kadhi

PARTICLES DON'T HAVE A GENDER, BUT IT'S TRUE, AS I HINTED in earlier chapters, that they don't obey the binaries we might expect in a world governed by prequantum conceptions of them. Quantum physics revolutionized the way that early twentieth-century physicists saw the whole world. The best way to get an intuitive sense of the phase transition that this induced in physics thought is to discuss the famed double-slit experiment. In this experiment, a metal plate with two slits in it is set up some distance away from a wall. Either a concentrated beam of light (like a laser) or a beam of elementary particles (say electrons) is aimed at the plate. In the case of the light, we perceive it to be like a wave—something bobbing up and down through space, kind of like waves in the ocean moving toward the beach. The light patterns that you see on the wall have the same shape that one might expect from a wave, as if it split and went through both slits. But, if you zoom in closely enough, there are individual, discrete dots on the wall, as if individual particles were hitting it, not a wave.

With a beam of electrons, which traditionally physicists thought of as little point-like dots, the shape of the patterns we see on the wall look similar—what one might expect if the electrons were waves, not particles. The pattern looks like each electron went through both slits, even though we know with a discrete particle, that's not possible. Even more weirdly, in both cases, if you attach a detector at each of the slits to look at what goes through them, the slits will detect individual particles, and the shape on the wall will be completely different from what we observe when there's no detector. What's going on? It turns out that light is made of particles called photons, and all particles, including electrons, exhibit wavelike behavior. Also, the act of observing the particles (such as by attaching detectors like I described above) changes what they do after we've looked at them.

Since discovering these fascinating features of particles, physicists have spent over one hundred years meticulously developing an understanding of the physical consequences. One result of that century of study is quantum mechanics, a core subject that all physics and astronomy undergraduates learn. Quantum field theory and the entire Standard Model of particle physics, which underpin the science described in the first two chapters of this book, are also consequences of this effort. We understand stars and a host of other phenomena in the universe because we've developed the machinery of quantum mechanics.

It's also the case that there's an ongoing dispute about how to interpret the findings of the double-split experiment, and research on this is still funded, although on an unfortunately limited basis. Solid state physics evolved as a field on the basis of a quantum understanding of materials, giving us many of the elements of modern-day computers and smart phones. These days, quantum information is the hot new kid on the block, and it's usurping a lot of the federal allocations that used to go to particle physics. As a species, we've committed to quantum mechanics, even though it's weird and completely unintuitive to those of us raised with a traditional Newtonian perspective on the world. We accept and work to intuit that particles like electrons are neither simply point-like, nor simply wavelike. Or, as Iraqi British drag queen Amrou Al-Kadhi explained to UK's Channel 4, they are nonbinary.

Yet scientists are among the people who have published whole screeds about the tyranny of trans and nonbinary gender identities and refuse to use the correct pronouns for their students. I opened this chapter with Al-Kadhi's comment because it highlights exactly how egregious this is, particularly when it comes from physicists: somehow we can do book-length quantum field theory calculations that are predicated on this nonbinary understanding of particles, but we can't commit to learning a couple more words when we learn someone's name or acknowledge that our categories are failing us socially.

In some sense, I understand how we got here. A lot of the conversation about gendered experiences in science is completely disconnected from gender theory. Scientists are not encouraged to pick up feminist studies literature. In one case (the "Sokal affair"), some scientists became famous for attacking gender and other areas of cultural studies as disciplines by sending a faked paper that trafficked in language developed in feminist theory to an academic cultural studies journal in the hopes of proving that a respectable publication would let even garbled work past peer review. Despite quantum mechanics offering maybe one of the best analogies for the complexities of gender identity available, physicists still often refuse to acknowledge a broad range of ways of knowing, including how feminist theory can enrich our understanding of what goes on when we do scientific research. Too many physicists also ignore the work of theorists like Karen Barad, who uses gender studies and quantum mechanics to inform each other in their book *Meeting the Universe Halfway*.

Indeed, I first met Barad's work not through physics but through my work in Black feminist science, technology, and society studies, where I spend a lot of time thinking about the experiences of Black women. Thinking about how physicists who are not Black women often don't listen to Black women about our experiences in physics helped me realize how biases are deeply embedded in the practice of physics. Our colleagues are quick to deny that a microaggression was in fact a reflection of any structural problems with sexism or racism; they are quick to question our capacity to evaluate the racial, sexual, and gender dynamics of the environments that we are in. We tend to think of these behaviors in terms of their impact on Black women or specifically on Black women's participation in physics. We complain that it is psychologically harmful and damaging to what science a woman does, and then we ask what science a woman might do if she wasn't experiencing exclusion.

A realization that it took years for me to come to was that the treatment of Black women can teach us a lot about the general practice of science—and the ways in which it is a deeply broken practice. White empiricism, a concept I developed for an article that appeared in the academic feminist studies publication *Signs: Journal of Women in Culture and Society*, is a practice of ignoring information about the real world that isn't considered to be valuable or specifically important to the physics community at large, which is oriented toward valuing the ideas and data that are produced by white men. In my paper, I developed this argument by using the specific example of how Black women are treated in the scientific workplace and juxtaposing it against a debate about whether actual observations and experiments are necessary to support theories of quantum gravity. Black women are constantly asked to provide hard evidence for our evaluations of our most commonplace experiences with discrimination, yet white men are taken seriously when they suggest that more affirming data isn't necessary in order to test their theories of quantum gravity. In other words, the way Black women are treated within the physics community is the canary in the coal mine for all women and gender minorities.

Empiricism is the ideological premise of the scientific method, the idea that meaningful information about the world is gathered through human senses: sight, touch, smell, sound, and taste.

Experiments are an empirical activity. I think this is a great way to gather information. The problem is that not all empirical data is equally accessible, because social asymmetries—differences—mean that we don't all have access to the same sensory experiences, and as Alexis Shotwell discusses in *Knowing Otherwise: Race, Gender, and Implicit Understanding*, there are other ways of knowing that aren't purely empirical. A white person can never know what it's like to be Black from a first-person perspective. This perspective, the person's standpoint, affects what data they can personally sense. You might assume that this has little to do with physics; after all, the law of gravitation has little relationship to what anyone thinks, no matter how much melanin is in their skin. But it's worth remembering that developments in physics are built collectively, through a human community that is constantly assessing the value of certain problems and making choices about who works for them.

Imagine, for instance, that a Black woman working at the boundary of what is known in physics has a fresh but controversial idea about it: do you think her idea will be taken as seriously as it would be if a white man said the same thing? In a context where Black women are disbelieved about our expertise, our ability to contribute to communal knowledge production is fundamentally at risk. At the same time, white men in string theory who argue that empiricism is no longer a good scientific standard because it doesn't support their favorite theory for quantum gravity are taken seriously and encouraged to explore the implications of their ideas. This is one side of white empiricism, where misogynoir—a term coined by Moya Bailey and expanded by Trudy to describe the specific practice of white supremacist patriarchy against Black women—shapes which concepts are taken seriously in the scientific community. White men can question empiricism entirely while Black women are questioned for practicing it at all. In other words, white supremacy operates in science as a form of anti-intellectual, anti-empiricism.

If the testimony of some people does not matter, it means that we are not evaluating evidence on its own terms but rather evaluating evidence on the terms of who the evidence comes from. Someone who enjoys playing devil's advocate might argue that our evaluation of evidence is always contingent on our level of trust for the person who provides it and that this is not inherently a bad thing. Yet we know the fundamental suspicion of Black people is violently dangerous and fundamentally racist, and it should be a given that it is also inherently bad to treat non-Black trans and genderqueer people with suspicion. The idea that some information is more suspicious because of who provides it is incredibly dangerous when the evaluation is based on ascribed identities.

Patriarchy, the system where men are a class of socially, politically, and physically dominant people, is what I call a total system: it touches, shapes, and ultimately distorts everything in our lives. I did not understand this as a young girl who was learning about sexism, and I didn't understand it as a young woman who was combating sexism either. I didn't understand that prejudice against women was rooted in a total system that, like and entangled with white supremacy, was creating what Melissa Harris-Perry calls "the crooked room" that I was trying to stand up straight in.

The evolution of this system began in early modern Europe, in the ideas of seventeenth-century English philosopher and doctor John Locke. In *Vexy Thing: On Gender and Liberation*, Imani Perry explains how the concept of "human" has historically been constructed specifically in relation to man (rather than woman): "The construct of 'man' for Locke was a relation of property. . . . Locke's definition was a relation that depended on sovereign authority. As Europeans traveled the globe, the sovereign authority granted them rights to property relations that were also effectively rights of conquest." In other words, the idea of a human was entangled with an early form of Manifest Destiny for men: from the point of view of the Europeans who eventually founded the United States and Canada, someone is human if they are a propertied European man, and if someone or something is a European man or a state representing them, then they have a legal right to all the property they can get their hands on. Patriarchy isn't just prejudice against women but instead a total system that says (white) men can own anything, including the capacity to determine other people's rights or even—in the instance of white empiricism—the ability to determine when evidence is required in determining reality.

Though I've read about patriarchy in books, it is my own experience as a Black woman and an agender person that allows me particular insight into how the total system of patriarchy works. I tend to think of why in terms of my favorite popular science fiction universe, *Star Trek*. In both the *Voyager* and *Enterprise* series, there are episodes where, due to quantum effects, the ship or some member of it ends up "out of phase" with the rest of spacetime. In the *Enterprise* episode "Vanishing Point," Ensign Hoshi Sato (the ship's communications officer) has an accident with *Trek*'s iconic teleportation device, the transporter. *Enterprise* is set soon after humans first develop the capacity to travel at light speed, so it is early in their use of the transporter technology. Without giving away too much of the episode, Sato finds herself in what must be a personal hell for someone in charge of comms, trying to communicate with everyone on the ship but unable to after she gets stuck in a quantum phase that is out of sync with the rest of the ship.

"Vanishing Point" is a difficult episode to watch, partly because it's a reminder of how few opportunities Korean-American actress Linda Park got to flex her skills on the show, but also because it is a kind of torture imaginable to most Black women and gender minorities to be completely invisible and never heard. But what's also interesting about the episode is that by being out of phase, Sato gets holistic insights into the operation of the ship, and the lives of its crew, that no one else on the ship has ever gotten. She alone gets to see what Captain Jonathan Archer is like when he is alone with his dog, Porthos. Sato has a unique standpoint on the ship—and I choose this word very carefully. One of the major contributions that feminist theory has made to our understanding of science is advancing standpoint theory—the idea that someone's particular social position can give unique insight into phenomena, insights others with different social positions would miss.

This is what my journey into understanding what it means to be an agender woman has been like. It has forced me to look at patriarchy from angles I never would have considered if I had maintained my belief that I fit neatly into the gender binary rather than recognizing that I am what Ralph De La Rosa calls a "gender dropout." Maintaining a traditional relationship to the gender binary required accepting, uncritically, what I had been told by society about who I was and who I could be. This too is a hallmark of patriarchy, and it is one that even cisgender women should be able to understand. Patriarchy says that people have to be certain things. It says that we all have to accept our social gender/sex assignment, based on our birth genitalia or the genitalia that are selected for the intersex among us at birth. It says that boys must grow up into domineering men; girls must grow up into refined ladies.

As Zakiyyah Iman Jackson puts it in *Becoming Human: Matter and Meaning in an Antiblack World*, "black womanhood is imagined to be a gender apart [and] also an 'other' sex." Indeed, one of the challenges of Euro-American discourses about trans, and more generally gendered, identities is that they are centered in traditionally Western framings of what constitutes gender, sex, and the numbers of genders and sexes out there. For those of us for whom white supremacist settler colonialism has been an imposition, the Euro-American discourse can be both incomplete and a hindrance. Whiteness has an intimate role in how people like me are able to understand our gender and sex identities. It's important to recognize too, as Marquis Bey points out in *Anarcho-Blackness: Notes Toward a Black Anarchism*, "The gender binary is part and parcel of capitalism's division and devaluation of gendered labor."

When I came out as agender, it was immediately obvious to me that my experience with "agender" was different from what a lot of trans people go through, and for this reason I've always felt a bit hesitant about my inclusion under the "trans" umbrella, even though I guess I live somewhere out there, on the outskirts. I had a lot of questions when I came out, and I continue to have a lot of questions now. There seem to be all of these rules about being trans or being on the periphery. If you feel fine with the so-called gendered pronouns that people use for you, are you really agender? If I continue to be femme presenting, doesn't that mean I have a gender? I trust that other people have strong gendered identifications, but I don't understand it. I kind of think that a strong gender identification is like quantum spin—you either have an internal sense of it (a spin greater than zero) or you don't (spin zero). Of course, this analogy is imperfect. I believe my understanding of how to articulate my identity will change with time, because our understanding of gender and sex is always in flux. But for now: I am genderless yet in my everyday life I am gendered by others. There is a distance between what people believe my gender is and how I feel on the inside. My subconscious gender experience does not align with my physical and social sex experience. Most of the time I am received socially as a woman, but occasionally my presence in a women's bathroom is questioned.

At the same time, I distinctly remember asking myself in my early twenties what exactly made me a woman. These questions were prompted by discourses about sexism in science that typically focused on white women. For a long time, I thought my only discomfort with them was because they ignored race. But I was feeling a discomfort with a conversation about patriarchy that only acknowledges women and usually means cissex women. The very existence of trans women and men, women and men who have a complex gender identity, and nonbinary people is regularly erased entirely from the discussion of gender, sex, and science. When trans women, trans men, and those who are somehow outside of the binary do become part of the discourse, it is usually in the most superficial ways. All trans people are often classed together as a group, even though trans women and men tend to have very different experiences with medicalization, socialization, transition, public gendering, and violence. Because these experiences play such a major role in the lives of trans people, we must integrate them into our understanding of how patriarchy affects science and particularly harms scientists who aren't cis men.

We know from the *LGBT Climate in Physics* report that trans and nonbinary people in physics are particularly harmed by gender discrimination, including by advisers and colleagues who refuse to use people's correct pronouns, a damaging and cruel practice known as misgendering. It should be obvious that when you refuse to respect someone's pronouns you are making a statement about what's important and what is not. First-year college physics students are expected in just one semester to not only memorize Newton's laws of physics but also to learn how to apply them. If we can have the lofty expectations that our students will master the basics of gravity—a deeply mysterious force that pervades the entire universe—then surely they are owed mastery by their professors and classmates of a couple of letters that get their pronouns right. To tell students that it is too difficult is an egregious, brazen lie.

When we choose to ignore or refuse to learn the pronouns of our students and colleagues and actively misgender them, we are reifying the practice of white empiricism in science and in doing so putting everything that matters to us at risk. As with the choice to dismiss the observations of white supremacist patriarchy by Black women, ignoring an individual's pronouns is a clear statement that you don't believe that misgendering reproduces violence. To say that you have a philosophical objection, as some faculty have taken to doing, is a form of cruelty. Who gets hurt when someone is called by a name that it is comfortable for them to answer to? Absolutely no one.

For all the ruckus about how "the millennials" are too sensitive, the people who insist on using "she" when someone knows they are better described by "he" or "they" are the ones who are weak and cannot get it together. Superficially, this discourse is amusing because the people whining about sensitive kids will label pretty much anyone younger than them as a millennial. I am a millennial, and I will be nearing forty when this book comes out; according to Pew, the youngest millennial will be about twenty-five. Odds are most of the time they mean Generation Z but likely they also don't actually care about the difference because care is not in their emotional vocabulary. But more deeply, the intensity of the reaction against respecting the pronouns and gender identities of trans and genderqueer people reveals a selfish proclivity to refuse to let humanity grow and to let its children breathe.

The challenges to respecting pronouns don't always come from the people you might naively expect them to come from. Trans-Exclusionary Radical "Feminists" (TERFs) believe that because of how we are publicly gendered, we have to accept that trans women have benefited from being publicly gendered as boys when they were children. Arguments like this can be found in fairly mainstream publications, and as a friend said when one such deeply transmisogynistic op-ed by Elinor Burkett came out in the *New York Times*, such arguments are a "trans-hating shit storm of awful retrograde fuckery." To explain why we think it is a THSSOARF, it's important to understand that trans people often (but not always) experience cognitive dissonance about their public gender, their socially assigned gender at birth, and their internal sense of gender, very early on. This is often experienced as harrowing and torturous. For trans people—and likely other agender, genderfluid, and genderqueer people—the consequences of feeling forced into the wrong part of a binary can be severe. Around that time, a trans friend told me that all of the trans women they knew of who'd died had died either via suicide or murder. The white trans women they knew committed suicide and the trans women of color were murdered. No one died of natural causes. It was the Burkett article that made me realize that I could not stay silent about my own experience with gender. It is what led me to write a blog entry entitled, "I, Chanda, Agender."

Importantly there is no single, universal trans experience. Some trans people have a strong internal sense of gender identity but have zero interest in or even desire to modify their bodies. For some trans people, reconstituting their body with medical support is a matter of life and death. All of these experiences are legitimate and should be attended to. For some trans people, being regularly misgendered as a young person ends up being a long-term trauma. For others, it is not. The fact that it is not for some of us does not erase the reality of what it's like for others. I tend to think of the existence of the genderqueer and trans communities as a challenge to the proclivity to homogenize human experiences with gender and sex.

Skin color and body shape also play a major role in how we are perceived. For example, I didn't know until Twitter became a thing that many white people thought that Missy Elliott, the genius rapper who is always dressed up like a high femme, was kind of butch and tomboyish. Her style is very feminine from my point of view. The claim that she's butch is deeply entangled with the way Black women, particularly those who are darker and bigger, are often thought of as less feminine—always measured against heterocisnormative and very white standards of what makes someone seem feminine.

Under the white gaze, the same one in which Sara Baartman was made into a zoo exhibit in the early 1800s because of her big butt and dark skin, Missy Elliott isn't enough of a woman because she's not slim or light-skinned and wears tracksuits sometimes. Beyond it, part of the damage that slavery has done to our community is that few of us are familiar with the structure of gender and sex in our home Indigenous communities because most of the Black Atlantic has no idea what our home Indigenous communities were. Even for folks who do know their history and origins, for example, people from Indigenous American nations, the continuing impact of colonialism is to make it difficult to claim inherited traditions, including around how gender and sex identities are understood.

Even staying in a Western context, my own efforts to describe myself reflect the difficulty we face with language. One way of describing the fact that I don't have an internal sense of gender but move through the world as a woman is to say that I am genderless, but that my sex in relation to culture is female. Beneath the substrate of these descriptions is a more complex reality. The bifurcation between gender and sex is artificial, yet in recent years describing my own sense of self seems to require clinging to some distinction between the two. The words represent different things to me, and at the same time, I know they represent wholly other concepts and experiences to people who are completely different from me. It's important to also recognize that academic consumption of identities can act as a form of disciplining and control. The words "transgender" and "transsexual" are often given specific meanings in academic contexts, but it is not only and not primarily paid intellectuals who have an investment in the direction the language goes in. Part of the task of scientists who work on the boundaries of gender and sex identity, and those of us working in women's and gender studies, is to understand the importance of non-academic discourse about gender variance.

While some of our trans activist intellectual heroes work day jobs in academia, most of them do not. Instead, some of our most important trans activists, like the late Blake Brockington, can be found out on the streets. By this I mean not just organizing protests, but sometimes also literally living on the streets. Trans youth, particularly trans youth of color, experience high rates of violence and homelessness. Trans women of color, especially Black trans women, have historically been some of our most important philosophers but also are murdered at disproportionate and frightening rates. Black trans men and the violence they face is too rarely discussed and acknowledged. While I was working on this book, Black trans women and men were both murdered by police, vigilantes, and intimate partners. Nearly all of them were under the age of thirty.

I do not want to mourn the loss to science these murders potentially represent. Instead, I simply want to make note of how white empiricist science—an umbrella term that I construct to include transmisogynistic and patriarchal science—shapes the discourse about who counts as human and the ways in which people are allowed to be human. I also want to make note of the dreams and thoughts that are precluded by this violence. There are trans nerds of color who may be as fascinated by spectra as I am; too few have the opportunity to become experts on them like I have. This is an injustice not to science, but to those who are denied an opportunity to walk the intellectual pathways that their brains might unfold. It is an injustice that impoverishes humanity on the whole, but those on the margins pay the highest price for it.

The lazy conclusion to come to here is that this is a problem that will be solved by more diversity in science, and that the inclusion of token people from minoritized groups would signal that the problem has been solved. But I'm not trying to make an argument for the sort of directionless, all-encompassing inclusion that is so often held up as a panacea to the problems of a backward scientific culture. The goal is not just to have more trans scientists, but to build a world where being trans isn't a barrier to full participation in society.

On the other hand, arguments for "inclusion" can sometimes be used to obscure violent realities. As Jin Haritaworn and C. Riley Snorton have asked, "What would a trans politics and theory look like that refuses . . . 'murderous inclusion'?" This question, from the article "Trans Necropolitics: A Transnational Reflection on Violence, Death, and the Trans of Color Afterlife," arises specifically in response to how discussions about the violent deaths of trans people of color operate as a kind of currency in academic diversity and inclusion discourse. The murders of trans people become a springboard for making a milquetoast point about being nicer to trans people at work, rather than radically restructuring society to protect trans lives. Haritaworn and Snorton raise a necessary question about demands for inclusion: "inclusion in what?" Dean Spade has written extensively against one form of inclusion, pinkwashing: "a term activists have coined for when countries engaged in terrible human rights violations promote themselves as 'gay friendly' to divert attention from terrible human rights violations, [for example] diverting attention from the brutal colonization of Palestine." More generally, Spade says, "The right is also leveraging trans issues as a tool for promoting right-wing security and military agendas."

The fact of pinkwashing and the specific centrality of trans inclusion in the military as a "progressive" issue while the military continues to engage in aggressive operations that put the lives of Afghani, Iraqi, and Central and North African trans and queer people at risk, makes clear that "inclusion in what?" is a persistently necessary question. The current scientific establishment, especially physics, has a strong relationship with totalitarian, racialized structures through, for example, our ties to military weapons development and other technologies of control. These structures are increasingly putting all of humanity at risk, but most especially gender transgressive people across the Global South—from North America's ghettoes to Barbados to Vietnam to South Africa. A science that proclaims to support trans and queer identities in the context of such violence is not subverting white empiricism, it is reinforcing it.

Drawing such broad connections might seem irrational, but if there is any lesson that ought to be learned from the catastrophic reality of global warming it is that we are not so disconnected from one another as liberal, individualist notions might have us imagine. The Cartesian bifurcation of subject and object is an attractive way of looking at the world that obscures some of its most basic features. If I celebrate American nationalism, and American nationalism is predicated on the erasure of the Indigenous nations of the lands that we call America, then I am celebrating a structure that demands the erasure of Indigenous people. If I celebrate womanhood in a way that excludes women of the trans experience, I have contributed to the discursive dehumanization of trans women. Because trans women of color deal with the intersection of dehumanization based on race, gender, and trans status, they are put most at risk by activities that compound dehumanization along any axes of their ascribed identities.

If white empiricism is an orientation that leads to both the silencing of Black women—the canaries in the coal mine—and all women and gender minorities as truth tellers, and to the corruption of a strong objectivity within science, then what is the solution? The anti-patriarchy solution might just be queerness. Joe Osmundson and I talk a lot about how, at its best, queerness is conceptually always at the edge of knowledge—a refrain we picked up from José Esteban Muñoz's *Cruising Utopia: The Then and There of Queer Futurity*. This queerness defies the proclivity of scientists to put people into neat, static categories because what constitutes queer and even what language we use to describe it is constantly shifting, advancing forward into the human unknown.

This is a multifaceted effort. There are biologists constantly learning new things about what underpins sex and gender, which are both biocultural phenomena. There are gender theorists who are questioning the gender theory of yesterday, of European colonialism in decolonial and postcolonial frameworks, ultimately producing new ways of looking at the culture we wake up in and choose to create tomorrow. Understanding that, as Peter Cava says in *Trans Bodies, Trans Selves*, "trans politics call for universal liberation" means recognizing the ways in which academia, which includes academic science, is bound up in ongoing injustices that form the barriers to collective liberation. As Marquis Bey says, "Socio-political gender transgression marks a distinctly anarchic practice."

In other words, we're here, we're queer, and like quantum mechanics, we're not going anywhere.*

---

* And, importantly: this chapter could never be a comprehensive and complete introduction to trans identity and the issues faced by gender-variant people. Because no book can speak for the whole community, I encourage readers to seek out a wide range of perspectives and texts and podcasts and films and shows and music and visual art that will help you better understand the wide range of experiences of trans, nonbinary, and agender folks.

TEN

WAGES FOR SCIENTIFIC HOUSEWORK

> Where women are concerned their labour appears to be a personal service outside of capital.
>
> —Selma James, "Sex, Race, and Class"

In our stories of scientific genius, we don't talk about something that we all know intuitively from a very young age: we do not function without food, without water. In reality, we are organic substances that require continuous sustenance. This is the biological outcome of the physical principle: there is no such thing as a perpetual motion machine. Energy that is dissipated has to be replaced. It also turns out that sanitation matters. Newton hid from the plague during what is known as his miracle year; had the plague found him anyway, that miracle year would be nothing to speak of. We also believe that it matters that children are cared for and loved. We consider it a crime if they are neglected, even though it doesn't seem to be terribly criminal for police and vigilantes to shoot them if they are Black. But in theory, raising children, including the children of scientists, is widely understood to be important.

Thus, before Einstein's equation, before the detection of gravitational waves and neutron stars, comes sustenance. Eating. The provision of eating. People who pick the food, feed the animals, slaughter the animals. People who deliver the food, cook the food, make sure we remember to eat it. And for all those pieces of paper we throw out, there are people who come to remove the mess, to ensure the office remains somewhat sanitary. People who scrub our toilets and mop our floors. Many of us, especially the cis men among us, have at points benefited from having someone at home whose full-time job is taking care of the cooking and in many cases the raising of the children. Einstein was not unique. Many men in science were like him. Someone else raised their children. It was, for the most part, their wives and mothers. Many women continue to raise the children of the many men who are still working in science. Even if those women have jobs outside the home, data shows that they are still doing more of the work inside the home.

We as a society don't talk too much about that, although the COVID-19 pandemic has forced some minimal discussion of it to the fore. We don't talk about the women at home making those Nobel Prizes and experiments and theories possible. The mothers and wives who kept sheets and clothing clean and did all the kid-related things so that their husbands could focus. We don't talk much about how this is work, and it is work that they are not paid for. We also don't talk about the women who did the workplace version of this, the administrative assistants who typed up dissertations in the decades before the predominant scientific word processing software LaTeX came along. The admins who almost certainly often had to tolerate sexual harassment and sometimes sexual assault. They were certainly not paid enough and in many cases have been eliminated from the workplace.

We don't talk about the poor and working-class people, often immigrants, frequently of Black, Latinx, Indigenous, and Asian heritage, who get paid below a living wage to keep the floors at universities and labs clean and the bathrooms tidy.

Rather, in our stories of scientific genius, scientists don't have to spend their time helping people, they think great thoughts about science. They execute great experiments. They draw great conclusions. They stumble along the way, but eventually they stumble into greatness. We don't talk about how without all of this, scientific work is not possible.

We don't talk about how, because of all this not talking, people of color, white nonbinary, and white women scientists are often treated like extensions of a multiperson help force that keeps white men's science going. We are asked to be patient while white cisgender men can be as strident as they want. We will not be paid for this patience. We are asked if we can handle the "people stuff" because we are "just so good at that!" We are not paid for this work.

We are told we are natural educators and natural carers, but we are rarely given the same messaging about our research abilities. We are the ones whose schedules seem like 24-7 office hours for colleagues and students to stop by and cry and kvetch in. We are not paid extra for this work. We are the most likely ones to show concern for people being marginalized. We take on other colleagues' students when those colleagues can't be bothered to properly mentor a student who is not like themselves. We are not paid for this work.

We are asked over and over to meet with women in science clubs and dole out advice that the institution is not providing for its women students. We are not paid for this work. We are the ones who note that those meetings are mostly white cisgender women and that there are no similar meetings for Black students or Native students or transgender students or anyone at those intersections. We are not paid for this work. We are the ones who are asked over and over on Twitter to "keep educating us because we need it." Because why learn when you can ask a Black person to repeatedly be your brain for you? For free? It's an American tradition after all.

If you believe the stories of scientific genius, you might then think that all of my time as a professor goes to researching and teaching, but no. An inordinate amount of time is also spent on what is called "service." We have a lot of jokes about this in academia, like, "Committees and task forces are where agendas go to die!" As Sharon Traweek wrote in her anthropological study of particle physicists *Beamtimes and Lifetimes: The World of High Energy Physicists*, physicists tend to think of themselves as "a culture of no culture." Physicists tend to think of physics as separate from the world. But it does have a culture, one that makes it common to be the only Black woman in a physics class and one of few globally. As I discussed in Chapter 8, I remember, viscerally, how this feels.

Because I know these experiences intimately and because of my identities, people who are one or more of them often reach out to me for support. I get calls for help via email, Facebook message, and Twitter message. Sometimes it's from a non-Black queer student or someone who cares for one, but more often than not it's from a Black undergraduate or graduate student. There was a time when I was the undergrad who wished I knew whom to email. Later I was the Black grad student who sent those emails. As a postdoctoral fellow and now professor, I became the person responding to the Bat-Signal, even though early on I wasn't really ready and still had (and have) my own barriers to deal with. At least as a professor I get paid to do this service. I was doing it long before I was getting a salary for it though.

Responding to the urgent crises of students who don't go to the university where I work is generally unrewarded work and often occurs at the expense of my physical and mental health, and sometimes time devoted to my professional success. Even so, I try. In response to many generic requests for advice, I began writing my blog series, "Surviving and Thriving Series for Underrepresented Minorities in Physics and Astronomy/STEM." When I think someone else is better suited to respond to the situation than I am, I try to quickly make a connection between the person in need of assistance and the right resource. It's great when I can triage, although I've learned the hard way that sometimes the people I'm helping don't understand that these are favors I am calling in for them. But often I am the specialist who these students need to see. So I schedule coffee meetings and phone calls and write lengthy emails. I carry these students and their concerns in my heart and my head. When they tell me horror stories, I stay strong as I listen. I dispense advice. I make contacts. I let them use my name to reach out to people. I go home and cry about what they are going through. I bitterly accept that academic promotion committees are in fact going to label me high-risk because it is known, unofficially, that I unapologetically spend time doing this work.

Universities make grand proclamations about the value of diversity and inclusion but rather than taking action to address the needs of marginalized students, they form committees, and the committees put out reports and then administrators hand-wring about how they wish they could follow the recommendations of the report, but there's simply no money/time/commitment. Committees are where universities go to avoid spending money on things that would serve the university's apparent mission: research and teaching that is practiced in an equitable and just environment. I walk around knowing what this implies: those students aren't worth the effort. They were never going to publish anything important, so helping them doesn't contribute to the productivity of science, doesn't count as one of my contributions. I carry the weight of this knowledge in ways that the men around me often don't. Earlier in my career, I was told to shut my door to those students in order to protect myself. I irritated people by saying I would never. It's not who I am. Even if I can't do everything needed, I will at least sit down and talk with the student if that seems like it will help.

The people who get most swept into this academic housework are people who aren't white cis men. Also unsurprisingly, the people most likely to win awards and generous praise for doing this work are white (men). We people of color and white trans/genderqueer/cis women and nonbinary people do the academic housework that helps keep academia running. We are not always paid for this work, and when we are, it's often not enough. And we are punished for it when it comes time to apply for jobs and tenure. We deserve better. Not because some of us have PhDs or are working toward some kind of graduate degree. We deserve better than this because we're human. We deserve better than this because every single person who does support work that keeps science going deserves respect. Yet, instead of providing resources accordingly, making sure administration positions are properly staffed and janitors are well-paid and that parents of any gender have access to affordable childcare, the community would rather continue to blather on about a meritocracy and how it will all be fixed if we just get more diverse students into graduate school.

But the racism and sexism in graduate school *are* a problem, and they exacerbate what is already a poorly structured environment. Graduate students and postdoctoral researchers are paid low salaries, face economic instability in the form of not being able to guess where their paycheck will come from in six months or—if they're lucky—three years, and are vulnerable to tyrannical (psychological and physical) abuse from supervisors (professors) on whom they are totally dependent for a recommendation before they can move into their next position. Those same supervisors can, in many circumstances, also fire people outside of hiring season, making it impossible for someone to get a new position and continue to have a career in the field.

Researchers from minoritized groups, including gender minorities and especially those of us of color, face an extraordinary burden in academic spaces. We are expected on paper to perform like our straight white male peers: to win competitive grants that are dependent on our scientific ideas; to write a comparable number of papers that are subsequently published in highly impactful research journals; to teach well and earn comparable student-teaching evaluations; and to do committee work. This is the job of a tenure-track professor; postdoctoral researchers in science are supposed to focus only on the research-related aspects of the job if they want to have any chance at getting a tenure-track position. But in practice, both faculty and postdoctoral researchers will often find themselves doing far more. We also find that minority students see us as a needed resource and will knock on our door or send us emails, asking for urgent help as well as existential help.

Those of us with teaching responsibility will also have to work harder at teaching our classes because we know of the extensive research that shows teaching evaluations are biased against people of color and white women, but that universities refuse to replace them with more useful evaluation metrics. When it comes to committees, postdocs and even graduate students from minoritized groups who should be free to just focus on research will be asked to sit on committees, particularly those focused on "diversity, inclusion, and equity." Faculty will be asked to join these committees on top of their regular committee assignments; within their regular committee assignments they will often find they are marginalized because they have a different perspective from the dominant culture people in the room. All of this extra workload has to happen in the same twenty-four-hour day that white men live in—that's just science—but with no relief on the distribution of duties.

The other problem is that it's insufficient to write on your job or grant application, "For the last fifteen years, I have done more emotional housework supporting minority people than almost anyone else in physics." Emotional housework does not count as scientific work, not unless you warehouse it in some kind of formal structure, like a mentoring program that will require more emotional housework and institutional resources to maintain. Often the people most likely to actually do the hard work and be good at it are the least likely to have access to those institutional resources. So, people like me end up cleaning up the messes of the people who have the right structures in place to formally claim they are doing emotional housework, but we can't get credit for it. It is better to lie about being good at these things on your grant application than to be a person who actually does them.

People who make this work a priority still don't get hired. The number of Black, Latinx, and Indigenous North American faculty in physics and astronomy PhD-granting departments barely changes, decade after decade. At this rate, it will take a century before parity is reached. That is totally unacceptable. It is unacceptable because it is lazy, perpetuating structural racism to appease people who don't need and don't feel comfortable with radical change. Institutions say they care about diversity, but they aren't willing to do the heavy lifting of changing cultures that keep minoritized people of color out. It is unacceptable because it leaves the students who depend on Black faculty out in the cold and sends the message that their pedagogical experiences and psychological health are simply not important. In other words, there is no substitute for hiring.

At one institution where I was a researcher, the provost told the scholars in my program, most of whom were distinguished visiting Black, Latinx, and Indigenous North American faculty who held tenure at other institutions, that he cared for diversity but not at the expense of excellence. It's incredible how common comments like this are given how brazenly racist they are, especially since far too many academics swear up and down that explicit racism is no longer a major problem in academia, just "implicit bias." Yet these comments explicitly suggest that minoritized candidates for faculty positions tend to be less qualified, less excellent. This is a piece of the other way that I am expected to compete with white male colleagues: they don't spend their careers hearing that who they are makes them fundamentally incompetent and less excellent. Yet, administrators at our top institutions have no problem wondering aloud about their minority colleagues: "Seems like they will contribute to diversity, does that mean they will compromise our excellence?"

That provost who was worried that diversity came at the expense of excellence revealed himself to be deeply invested in a practice rooted in the mythology that white people—especially white men—are more likely to represent the excellence necessary to advance science than anyone else. People of color, especially those of us who aren't from certain Asian ethnic communities (e.g., typically academically successful ones like Taiwanese Americans), are a risk to science and therefore to progress and success. But even a superficial rummage through the history of science shows that people of color have been making contributions to "modern" science through the entire time period that we tend to call "modern." Rather than hinder progress, we have directly added to it, despite the very best efforts of people who could be categorized as "white" to stop us.

Is science really just white guys sitting around being excellent? Is that really how it happens? The white supremacist patriarchal mythology that underpins scientists' beliefs about how science happens is deep. Even though it is possible to overcome barriers erected by white supremacist patriarchy, I have found it hard to be at my best, scientifically, in this context of precarity. Yet, to be in the positions I held was also an opportunity. I got to spend my days fitting in thoughts about the origins of the universe in between worries about workplace racism and sexism, rather than having to do manual labor that I would find to be less enjoyable, in between worries about workplace racism and sexism. This difference creates class stratifications between poor scientists and other poor people. It means that when we talk about racism and sexism in science, we can focus only on people like me who are trying to access the property of "scientist." It means that when we talk about hidden and ignored contributions to science, the only hidden figures we go looking for are people who are members of the intellectual classes.

But people who have been pushed to the margins in professional, formal scientific settings have very often been making contributions through the labor that tends to be valued poorly, if at all, because it's not formally recognized as "science." The women (among others) who did caring work, raising the children of scientists—raising the children who would go on to become scientists—have been making a contribution to science this entire time. The people who cooked and cleaned for scientists contributed to science. The enslaved people whose labor provided the tax base for governments that then funded scientific missions were also contributing to science. Progress in science happens not just because of the scientists in the room but because of how their presence in the room is made possible.

It is rare to see people in the sciences actively link arms with poor, working-class, and other marginalized people in a way that challenges the social class boundaries between us. It is common for power relations to divide us. For example, those of us who somehow exist within the category of women, or a category adjacent to it such as femme men, tend to do more physical and emotional housework than those outside of that class. It doesn't have to be this way. In practical terms, rectifying this means reconsidering the labor-based relationships we have with each other that have traditionally been gendered. Indeed, I am not particularly interested in becoming part of the colonial establishment—the United States of America—and more concerned with how we can build communities that are in better relations with each other. We will all be better off when we value all of this unwaged work that people do at home and the low-waged work that people do in and around our offices and labs and provide the necessary resources to successfully do this work. This idea does not originate with me. Wages for housework as a concept was articulated in 1972 by my grandmother Selma James, who went on to cofound the International Wages for Housework campaign. This campaign proposed that paid workforce labor was made possible by unpaid household labor that, especially at the time, was largely done by women.

The concept of wages for housework may seem simple, but the implications are vast. For example, Black Women for Wages for Housework (BWWFH), which was founded in 1974 by Wilmette Brown and my mother, Margaret Prescod, demanded not just wages for housework and rights for women on welfare but also reparations for slavery and colonialism, ideas that in 2021 are coming of age in popular social justice movements. BWWFH's ideas informed a flyer put out in the mid-70s by the New York Wages for Housework Committee, which said, "As long as we have no money of our own because we work for nothing at home and for crumbs outside the home, none of us can choose whether or not to have children, and all of us face sterilization even if our tubes are not cut. Many women today refuse to have children because having a second job is the only way of getting some money of our own, and we know that we can't handle both and must 'choose' between the two. This too is forced sterilization." Black, Indigenous, and Latinx women on welfare were the primary victims of forced sterilization.

Bizarrely, mainstream feminists believed that an adequate response to patriarchy was not to protect or even expand welfare rights—and the bodily integrity of Black women—but to argue that women simply needed to work outside the home and be included in higher paying and more powerful professions. Earning the wages that men earn, many feminists believed, would equalize everything. Fifty years later, we know that these second-wave feminists were wrong. Millennials like me are choosing not to have children for a variety of reasons, including extensive wage growth suppression that has occurred in tandem with widespread entry of women into the white-collar workforce. Poor women who may or may not have waged jobs have also faced extensive cuts to welfare, which is the closest thing to date we have ever had to wages for housework, and in addition to unwaged caring work, many of them are working two or even three waged jobs just to make ends meet. Even those of us who make comfortable salaries often still face long work hours that make it difficult to manage the second job of raising a family. Wage suppression today is so bad that many full-time workers, including an increasing number of college faculty, are also dependent on welfare, which was extensively scaled back under Democratic President Bill Clinton.

I AM UP EARLY, WRITING WHAT WILL BE CLOSE TO THE FINAL draft of this chapter at my dining table in the Munger Physics Residence of the UC Santa Barbara Kavli Institute for Theoretical

Physics (KITP), where I am in week five of a six-week visit. My apartment is lush, replete with the latest anti-climate change environmental control technologies. The visit has been idyllic, the best research time of my life. Somewhere in the middle of the trip, I had the thought, "So this is what it is like to feel fully included and actually enjoy science." This is the product of careful decisions by the organizers, who chose a good group of scientists for the workshop. But also, ultimately, I have to thank the administrative and maintenance staff at KITP and the Munger Residence, who keep everything running smoothly and who generously worked with me to make sure I had what I needed to feel safe there in the wake of death threats I was receiving around the time I arrived.

During my first two years as an MLK Fellow at MIT, I was the only Black woman who worked for the physics department in Building 37 who wasn't a member of the janitorial staff. (And after I moved to Building 6, there were probably none.) But that doesn't mean I was the only Black woman in the building who contributed to the science that happened. The women who cleaned our offices, one from Jamaica and one from Barbados, were essential. They made sure we came into offices that didn't smell like garbage and kept our floors and bathrooms clean. For me as a Black woman physicist, they were the colleagues I felt most affinity with. They took a particular interest in me, especially when I mentioned that my family was from Barbados and Trinidad. They made sure I knew they were proud of me, and they regularly asked after my mother, whom they had never met or seen. When I talked to them about my grief at losing my grandmother the year before, they showed up with codfish cakes and bakes for me, the foods I most strongly associated with my Grandma Elsa, the foods I missed the most. No one else at work had that kind of care for my spirit.

These women, grandmothers themselves, likely understood better than I did the challenges I was facing in an environment full of white men who didn't understand where I came from and didn't care to. Even as one of them suffered from cancer, those two women came to work every day and laughed with each other and laughed with me. It was incredible emotional labor, and it is part of why I am a professor now. I often felt helpless to return the favor, but I regularly asked if MIT was treating them right, paying them enough.

Making science requires not just scientists doing research, it requires emotional housework to support (marginalized) students and scientists, it requires administrative housework from administrative assistants to make sure scientists can have colloquia and seminars and travel to conferences and committee meetings and visiting collaborators, it requires cleaning staff and dining hall/cafeteria staff and facilities/maintenance staff and IT support. *All* of these people who do *all* of this work make science happen. What would it mean to understand all of this work as part of the scientific project?

Recognizing and valuing that housework requires scientists to think of the humanity of the people who are not scientists but who support the scientific endeavor. It means acknowledging that it is our responsibility to support our administrative, janitorial, and food service staffs when they go on strike to demand better conditions and better wages. It means I will never regret barely passing my first quantum mechanics and differential equations courses my sophomore year of college. I barely passed my exams because I was fighting for living wages for Harvard's janitors because being a Black student and seeing that most of the Black employees at the university were being treated miserably sent a message to me about how Harvard valued people who looked like my aunts, uncles, and cousins. While (according to his own memoir) Pete Buttigieg quite comfortably ignored the twenty-one-day occupation of Massachusetts Hall by the Harvard Living Wage Campaign, I chose instead to join the fight to make Harvard recognize their contributions and the role they play in academic work, helping to coordinate food for the sit-in and manage the tent city that formed outside. Together with the workers, we won higher wages for the custodial staff.

Buttigieg ran for president in 2020 as a Democrat, arguing that he uniquely cared for forgotten working-class workers. Yet in his 2019 memoir, he argues that billionaire Facebook creator Mark Zuckerberg was across campus doing far more impactful work than us. Notably, Zuckerberg was still in high school during the sit-in, but more importantly, Buttigieg's attitude symbolizes the problem we face in valuing what is too often called "unskilled" labor because his views about it are typical. Understanding the contributions of janitors, dining hall workers, and administrative assistants means understanding that management styles that cut costs by cutting salaries through outsourcing or cutting positions are actually damaging science. It means understanding that when my former employer, the University of Washington, stopped paying people to empty the trash in our offices, they were cutting back on science.

It means understanding that part of science is emptying the garbage.

ELEVEN

RAPE IS PART OF THIS SCIENTIFIC STORY

Content warning: This chapter covers difficult topics, including sexual assault.

DISCUSSIONS OF SEXUAL HARASSMENT, ASSAULT, AND GENDER discrimination shouldn't be at home in a book that begins with wonder at the universe's majesty. They shouldn't have a home anywhere except history books. But a careful look at how physics works—as a social phenomenon—means talking about the fact that the scientific community may need a trigger warning. Sexual harassment in science and math begins early. It happens to girls who aren't even teenagers yet. Sometimes worse is happening to them at home or elsewhere. Even so, I wasn't sure I should include this chapter because I worried that media coverage will only focus on this chapter rather than the whole book, I don't want to scare girls (who are not the only victims but also the most likely victims of rape culture), and I'm more than my experience with rape.

I am also the woman whose interest in science comes with a flourish of utopian scientism. As a teenager, I believed that if we solved the fundamental equations of physics, the rest of the proper order of the universe could be derived. I've grown up, and I know now that science is inextricably tied to power. A friend once told me he feels consoled knowing that the universe is always getting its calculations correct, even if we can't understand them. But our quest to understand them, the very desire to, is deeply rooted in our relationship with power. We don't just want to explain events in the past like cosmic inflation. We also want to predict events, as discussed in Katie Mack's (delightful) *The End of Everything (Astrophysically Speaking)*, which details different ways the universe might die. To know the potential fate of the whole universe is to feel powerful. I wrestled with power—my own as well as that of others—while writing this book. In science, in academia, in capitalism, there are a lot of opportunities for people in positions of power to abuse others. It would be dishonest for me to not talk about one of the ways that this happens. But even so, I wasn't planning to. I didn't want to.

However, patriarchy and the particular forms of violence that go along with it damage our ability to relate to ourselves, each other, and the world around us. Rape is part of science, and a book that tells the truth about science would be a lie if I were to leave out this chapter. Whatever your response is to what follows, or if you choose not to read it, here's what I want survivors to know. Rape is an incredibly terrible thing, and we must fight like hell to eviscerate rape culture. Sexual violence doesn't happen to everyone, but it happens to too many people. Sexual violence is a real trauma, but it does not have to be the end or the biggest part of any survivor's story. It will always be part of what made me, but it will never define who I am.

I ACCIDENTALLY STARTED THIS CHAPTER IN THE MIDDLE OF writing another one. I was discussing my relationship to cosmic acceleration when I got to the part where I started graduate school, and the writing went downhill from there. Here is what I wrote:

> I was raped during my second term, and frankly, that was sufficiently distracting to knock the wind out of my scientific sails. It's actually pretty hard to focus when you feel disgusted with yourself and your life has stopped making sense. It's also hard to stay deeply connected to science when a scientist violates you so intimately. Plus, I couldn't afford (financially or emotionally) to be sent home, which meant I had to keep up with my classes. The nuclear physics that underpins how stars work is actually fairly complicated and kept me relatively busy, when I wasn't out trying to drink and dance my cares away.
>
> The thing is, though, that while rape changed my life, it didn't change me in some fundamental way. It took years to climb out of the hole that I fell into during that winter quarter, but as a postdoc, I found that worrying was not just a quirk of the cosmic acceleration problem but in fact a feature of how I do science.

Then I wrote that I didn't want to write about being raped in the middle of a chapter about the dark universe. No one wants to read about rape in a chapter about how cool particle physics is. But it begged a question: how is rape part of my scientific story? Because it clearly is. And then I found I was writing this chapter even though actually I wanted to write the ones I had planned instead. I couldn't stop writing this chapter, maybe because it has been growing inside of me for over fifteen years now.

So, how is rape part of my disordered cosmos?

There are a lot of spaces that I can't enter without being confronted with my rapist's name, my rapist's work, or my rapist himself. In between, there are the ways he tries to gather information about my activities relative to him. I hear about those too. But the context in which I do science goes beyond external professional networks and environments. It begins in my brain. It is surrounded by the other things that are happening in my brain. Like recently, I've also been trying to figure out how to look at my body without hearing the things he said about me, without having my concept of "am I sexy?" determined by how surprised he sounded when he told me he found me sexy.

In thinking about this, I've realized that the act of finding someone sexy or not sexy isn't really just about feelings, is it? It's also a power move. A power move where you tell someone whether they are sexually valid or not. Anti-rape activists say that rape is about power, and I guess I've always felt some conflict about this because it seemed very real to me in that moment that the pressure to take back my "no" was about his pleasure, not exerting power over me. But is it because of the power? Did he pressure me to take my "no" back so that he could feel powerful? Are men really so fragile that they can't control themselves? It's in the context of these thoughts that I do science.

I think about what else is out of control. The expansion of the universe, for one. It may be totally out of control. We don't know why it's accelerating, and we don't know if anything will shut it down. But maybe that's not out of control—it's just that the story behind it is beyond our control to know for now. In fact, science, as formulated in the centuries after the Enlightenment, is about trying to exert some control over the world by knowing more and more about how it works, then organizing information around rules and laws that can be used to predict what will happen next. In this way, we subsume nature, wrestle its unpredictability to the ground, and extend the natural limits of the bodies we are birthed with. Science has become a practice of control.

We judge and control, too, the way we practice science and the data we accept. Science is a collection of activities that are loosely organized around something we call the scientific method, but even what constitutes the scientific method varies from field to field. Some would say that science is about data, whether predictions from equations or experimentalists' data. All of this seems like it has nothing to do with power, until you ask: who has the power of observation? Whose observations are taken seriously? Who is deemed to be a competent observer? We judge each other's data and conclusions—is it always based on the evidence? Whose scientific method?

Who would believe me if I told my story publicly? (Which I feel I need to reassure my rapist, in case he is reading this, that I won't do.) The conversation around rape always seems to get hung up on one thing: he said/she said. What if she's lying, they say. Statistics show she almost certainly isn't, we say. Are you going to ruin a man's life because of statistics? They want to know. They don't ask themselves about all of the times women's lives, trans and cis men's lives, and nonbinary people's lives were ruined by men, just because. No statistics, just patriarchy and misogyny.

Seventeenth-century physicist Robert Boyle did not believe that women were competent as observers. In fact, he felt that the role of women was fundamentally to be dangerous sexual objects and that the best thing women could do for humanity was be as chaste as possible. We physics students are taught about Boyle's Law, if not in our freshman physics courses, then when we take statistical mechanics and thermodynamics. Boyle was the one who figured out that the pressure of a gas is inversely proportional to volume: shrink the space a gas is in, and the pressure goes up. Boyle's Law is very famous. His misogyny is not so widely known. He gets a pass because he came up with a physical law that ended up being crucial to the industrial revolution. He gets to be remembered well.

I don't share the identity of my rapist partly because I am not invested in being punitive, but mostly I am not interested in the punitive measures that would be taken against me: I would be disbelieved and remembered more for this one thing than for my myriad other contributions. The only witnesses who heard me say "no" are him and me, and I am certain that he's conscientiously discarded any remnants of that particularly inconvenient memory. Rape victims don't get to be remembered well, but rather get to be remembered as liars, bitches, and sluts who are now just ashamed that they were sluts.

Throughout Western history, men have always come up with good reasons for not believing women and throughout Western history many women have cosigned this. In some cultures, women who didn't cosign patriarchy were burned and hanged. Matriarchies were relatively uncommon but did exist. Western imperialists were not fans, though. Well, they weren't fans of the societies, but they were fans of those people's land, and they used incredible power and violence to take it.

And what is the relationship between power and why we do things? I wanted to control nature by controlling cancer, once upon a time. Then I wanted to control nature by knowing it intimately.

How is rape part of my scientific story, or how is science not like rape when both are about power? Both are indeed about power, but this is not manifest to us, which is part of how the power dynamic works. We are told that rape is about desire when really it is about ownership; we are told science is driven by curiosity when really . . . I still can't bring myself to say it is about control because I'm not sure it always is or has to be. But that's what it has been when we integrate all of the scientific activity in the last five hundred years of the Western world. That's what it is.

Rape is part of my scientific story because every day, at least once a day, I think about rape. I think about how I was raped. I question whether it is OK to call what happened to me "rape." I think about the fact that many women are raped, and most women experience sexual harassment of some kind at some point in their lives. I think about what he said to me when my clothes came off: "You know, you actually have a cute body there." He was surprised. He was giving me drinks all night and then marijuana and he was surprised my body was cute. Why was he planning to fuck me if he didn't feel that way from the start? When I was that age, I looked young. I looked potentially underage.

Rape is part of my scientific story because every day I think about rape and every day I am striving to be a scientist and every moment that I am thinking about rape, time is being shaved off my life. I am being accelerated toward a higher blood pressure and heart disease—the things the Black women in my family have been unable to avoid. A woman in my family was raped years before I was born. She called me one Friday and told me this fact in the middle of my workday—a day when I was doing science because I am lucky enough to have this job.

Rape is also part of my scientific story because during my first semester as a professor I left the physics faculty retreat for a couple of hours to go to a talk on sexual misconduct in STEM where "sexual coercion" was mentioned a couple of times as an example of the type of sexual misconduct that occurs in STEM. After the talk was over, I casually discussed being raped with a couple of the women faculty who had also been raped and then a couple of the Black men faculty talked about how it turns out fellow Black astronomer Neil deGrasse Tyson may have raped someone. I had written a *Scientific American* piece about that, and that's the thing: my first *Scientific American* byline was about rape. After the piece was published, women wrote to tell me stories of things that happened with Neil deGrasse Tyson that made them uncomfortable and also to share stories of rape and sexual misconduct by other perpetrators whose names I do not know.

Rape is in my inbox pretty regularly, and also my social media timeline. Maybe rape is part of everyone's scientific story except that some people get to treat it like a footnote while the rest of us have to live with whole chapter titles, or even reckon with whole long books about the fact of rape: it happened to me, it happened to my friend, it happened to my elder, I'm unsure if it has happened to my sister but I worry because there is always a future.

To the editors: rape forms a through line in my story. It is a discursive timeline. Rape is part of this scientific story because about fifteen minutes before I wrote this paragraph, I was writing about Bose-Einstein condensate dark matter, and now I am writing about rape, because to write about science and gender is to inevitably, at some point, write about rape because rape happens in science, all the time. I was raped at a science conference by a scientist. I was once also followed into my hotel room by a scientist and pinned against my hotel room wall and kissed without my permission. At the time consent didn't factor into my thoughts. I thought that this is just how some men are, and it's uncomfortable and why did I let him follow me to my room. It didn't occur to me that he never asked if I wanted him there, and it's only become clear to me in the aftermath that he took my disinterest as a source of encouragement. He believed it would be romantic to follow me and pin me to my wall without my consent and impress upon me with his lips how perfectly we went together.

"Rape Is Part of This Scientific Story" is a chapter that nobody wants to read in a book that maybe people were thinking about finishing. Rape is the chapter in my life that I didn't want to live in a life I was thinking about actually living, but one I thought at times about ending early. After the rape, living felt effortful in new ways. I drank a decent amount and still learned enough quantum mechanics and nuclear astrophysics of stars to get by academically, but inside of me I couldn't get the knowledge of what had happened to me out of me and I wanted to swallow so, so many pills.

This is the chapter that continues to unspool in my brain. I want to let it go, but it stays with me, night after night. I come back to it. I leave the file open and then I think, what should I write about, and it says, "You want to write about how rape is part of your scientific story." And I think, "Nobody wants to fucking write about how rape is part of their scientific story." But this is what rape has been for me. It became a part of my identity. Raped. Person who has been raped. Student who was raped by a more senior person. Postdoc who was raped while she was a grad student. Professor who was raped while she was a grad student. Graduate student, postdoc, and professor who sometimes has panic attacks during sex. I am a professional scientist whose entire job is about directing her mind, and I am unable to do this basic thing of directing my mind past rape.

This is me: spending over a decade trying to assimilate the idea that I am a rape victim. I am supposed to say survivor. Sometimes I do. That is my identity too: person who is trying to accept this/person who does not want to accept this/person who revises each time. Maybe it was consensual and I'm making a big deal out of nothing, I think. For the first two weeks, I didn't remember saying "no." I just knew I wanted to die, drive my car off a cliff, those things. But then I remembered saying "no," and it is one of the worst memories I have, right up there with finding out my most beloved, my grandfather, died and having my jaw broken in two places by a car driven by a man who was too worried about his taxi fare to worry about whether there were people in a bike lane. Being raped is, in some ways, worse than having a jaw that is broken so badly that wiring it won't fix it and several hours of surgery are required to install titanium plates. It is worse than all of the subsequent root canals, the two rounds of braces, living for a year with ten fractured teeth with their nerves exposed, and tearing up nearly every time I ate or drank.

Maybe at this point I don't have anything else to say about how rape is part of my scientific story except that I didn't actually feel nauseated until I started to say this, and it's because it's a lie that I don't have more to say. I have a lot more to say, but there are things I'm afraid to say because I'm afraid of my rapist, I'm afraid of the people who want to know who he is, I'm afraid of the people who want to know who he is and also are committed to not believing it was rape, and I'm afraid that I will believe them. Each time I remember that I said "no" it's like an out of body experience. If that did happen then why did what happened after I said it happen? When I was hit by the car I thought, as I went down, "This will hurt in the morning." I thought I was going to be sore. I saw the future.

When I said "no," I didn't envision a future. There was no future while he experienced pleasure and I waited for him to be done. There was the window where I could see a white sedan, maybe a Nissan, under a sodium street lamp. Five hundred eighty-nine nanometers, my favorite electromagnetic frequency, for no good reason except it was the first one I learned. I don't remember how I got back to my room. I just remember I called my boyfriend—from the plane?—and broke up with him. I remember I felt filthy. I remember I believed myself defiled and irredeemable.

I don't remember doing homework the subsequent week, but I know I did. I passed my classes. I knew enough to pass my qualifier the following year. I was filthy, and I was a PhD student in one of the most renowned astronomy programs on the planet. They were both somehow true, empirically.

Rape is now part of my story of being a professor of physics. In week three of being a professor, I woke up one morning and saw three different stories about three different rapes in my Twitter newsfeed over the course of ten minutes. As I think through the stories I am reading, I cycle through this: first, maybe she is a liar. Then: maybe the police are gleeful at arresting a Black man because Lord knows they are racist as fuck. But maybe the police are gleeful at arresting a Black man because Lord knows they are racist as fuck and but maybe he's a rapist and so far criminal justice is the only way we as a human community have decided to collectively confront the act of rape.

But also I hover around maybe she is a liar. I hate myself for having this thought because I know it is the thought that people would have about me, and it is the thought that terrifies me the most because it is the one that would be articulated and used to smother my internet search results. I would be remembered for that and not any of the other things I did with my life, my life subsumed by one of the worst things that ever happened to me. I don't want to be the one who really becomes famous because she was raped by ______.

I wander away from writing to check email and Twitter because the thought of people finding out ______'s name is so horrifying that I have to stop thinking.

But on Twitter it's just lots of people talking about misogynoir. There in my email inbox is a thank you for a donation to a Jewish organization that is being attacked for recognizing the humanity of Palestinian people. *Underneath all of this lurks rape*, I think. The rape that Palestinian women have surely had to deal with. The rape that Jewish women in the Israel Defense Forces have also had to deal with while they were violently policing the boundaries of the Palestinian women's lives. The rape Black women in academia have to deal with. I am a Black woman in academia. The rape I have to deal with. Forever. Rape is part of my scientific story because I am forever going to be a woman in science who was raped, and we can't and shouldn't stop talking about all of the rape that is everywhere because the rape keeps happening because there are people who . . . well I don't know why the rape keeps happening. Except that, as I said before, anti-rape activists tell me it's about power.

In physics, power has a very specific meaning: it is the rate of changing energy, of doing work. Work has a very specific meaning too: work is when a force causes something to move along the direction of the force. Like say, when a man forces you to be close to him, he is doing work. The object being moved does not do any work. Power is the amount of work that the man does over the time that he is forcing you to be close to him. The object being moved does not have any power associated with it.

I want to have the power to eject this memory: to force it far, far away from me. By that I mean I would like to have the power to eject this memory into the nuclear inferno that is our sun. The sun is, effectively, a series of nuclear explosions, mostly converting hydrogen into helium. Better this memory blow up inside the sun than inside of me. But this memory is written on my body so instead I have to trace the lines of force that are available to me. I look to see what work is possible. For years, I had nightlong knife fights where I was the only person present.

Right now I cannot see much beyond the fact that I can move people with my words. I can work to pull the pieces of my story together and I can send that story out into the wider world, hoping for the best for the people it touches. Lacy M. Johnson wrote in *The Reckonings*, "I want him to spend the rest of his life in service to other people's joy." I want that too, but I don't have any hope for it. I know he will seek the spotlight and power as he pleases, never holding himself back, never thinking about his responsibility to transform the conditions in which I work. I know the only force I can possibly enact on him is to allow myself to become a nuclear inferno, engulfing him, yes, but also converting myself into something else: no longer a scientist, just that woman who a lot of people think lied about that guy because she hates men and wants attention and well, her science wasn't going well so she decided to do this. Helium is lighter than air; to be a rape victim who publicly accuses her perpetrator is to become unmanageably heavy.

The sun gives out 38,460 septillion joules of energy per second while it converts hydrogen to helium, which means its nuclear explosive force is the equivalent of ninety quadrillion tons of TNT per second. One day, the sun will run out of hydrogen, unable to sustain these explosions, and it will die. The resulting planetary nebula will look enormously beautiful from far away. But close in, it will destroy the solar system. The explosion looks beautiful from the outside, but inside, a life-sustaining planet will be dying.

I work at ways to cope, to be life self-sustaining. Each time my diagnosed post-traumatic stress disorder (PTSD) threatens my day, my week really, my coping mechanisms shift. On the first day, I usually struggle to do more than sit. Thinking hard things, doing calculations, is almost out of the question. On one of those first days I cried on the phone three different times, including on a call that was ostensibly related to work. The call was for a small group that is committed to changing how astronomers relate to the wider world and each other; the woman and nonbinary person listened, let me cry, and reassured me that if I needed to cry on that call then letting me do it was part of the work.

Another time, I put four days into writing a book review about Jonathan Kahn's *Race on the Brain: What Implicit Bias Gets Wrong*

About the Struggle for Racial Justice. The essay I produced, "Diversity Is a Dangerous Set-up," is one of my most popular blog posts. A friend printed it and posted it outside her office in an Ivy League chemistry department. Another friend turned it into a zine.

The time I am writing from right now is one where I decided that I wanted to know about other people's experiences, rather than just fight. I am experiencing a need to be validated, especially after some disappointing conversations with men in my life. My spouse, who does everything he can to take care of me during weeks like this, is traveling for work and can only soothe from afar. The night before he left, I dreamed he extended his tallit—a Jewish ceremonial shawl—over my shoulders and held me tighter than he did when I entered his tallit during our wedding ceremony.

While he is gone, I will imagine myself safely wrapped up with him. But I also must invent other ways to cope. I've thought about compiling a survivor kit—one with the warning that we each survive in our own way, and this playlist/reading list/activity list is just my surviving style for this week. If I wrote a blog post about this, it would go something like this: I ordered Frances Driscoll's collection, *The Rape Poems*, and stayed up late one night reading through a third of the book. I also called a friend who is both a survivor and an academic expert on rape and technology. I called her specifically to ask her if my reactions to widespread media reports about a rapist are normal. She reassured me they are. She also told me that the guy who gave me a hard time for talking about it can go to hell. The problem with him going to hell, though, is that I love him, fiercely.

I also cope by circulating articles about a new documentary that has just premiered at a film festival, detailing the experiences of Black women with a famous Black man rapist. A famous Black woman producer pulled support for the film; rumor has it that it's because she didn't want to seem like she was coming after Black men. This begs the question of why no one pressures Black men to avoid seeming like they are coming after us, after Black women. Someone posted on Facebook that she watched the Kobe Bryant case unfold at age eleven, already a victim, learning what would happen if she talked about what had happened to her. So many of us learned from that case that beyond actually being raped, not only was being a rape victim bad news, but if you're going to be raped, try not to be raped by a talented Black man who Black people adore, because we live in a world where Black people already have to struggle so much. But also try not to be raped by a mediocre white man because in all likelihood, he will get away with it. The hints that he will get away with it are frequent. Christine Blasey Ford calmly testified to the whole planet that Brett Kavanaugh had assaulted her. Testifying altered her life, presumably for the worse. Brett Kavanaugh made it all the way to the Supreme Court. He denied it, weeping tears of rage at the allegation that he did something that is so common it is almost mundane.

Rape is now part of my struggle, as a person and as a scientist. Part of my survival mechanism has been by adding to this chapter, which I both fear and understand will never really have an ending. I started writing it after I saw my rapist at a party. Now, I look to my right and see equations that I wrote down this morning. Today is day three of this latest PTSD flare-up. I feel like a failure. White men who weren't raped and don't have to deal with racism have been more productive than me for the last three days. But also I did write down those equations and the impetus for doing so came from conversations I had over the last two days. Somehow, I did things. This chapter has no ending, but I'm still going. I can only conclude that I still want a chance to be not just a rape victim but a rape victim who chooses life.

PHASE 4

ALL OUR GALACTIC RELATIONS

Knowing that physics is a human phenomenon, we now try to understand:

> can we situate ourselves, collectively and humanely, in the universe?

TWELVE

THE POINT OF SCIENCE: LESSONS FROM THE MAUNA

Our relations to each other, our prayers whispered across generations to our relatives, are what bind our cultures together. The protection, teachings, and gifts of our relatives have for generations preserved our families.

—Winona LaDuke, *All Our Relations: Native Struggles for Land and Life*

In fall 2011, I arrived at MIT a newly minted Rev. Dr. Martin Luther King Jr. Postdoctoral Fellow in the Department of Physics. I had left my fellowship at NASA after it became clear my planned research program wouldn't work out there, but before I left, I met MIT Professor Paul Schechter. Astronomers will instantly recognize Paul's name because the Press-Schechter formula, named for Paul and William Press, is one of the most important equations in galactic astrophysics. It predicts how many galaxies we are likely to observe in any given scoop of spacetime. Paul is something of a giant in the field of astronomy and for reasons that remain a mystery to me, he took an interest in me. Right after I arrived at MIT, he insisted that I go on an observing run—a trip to a telescope to collect data—with him and a graduate student at the twin 6.5 meter Magellan Telescopes at Las Campanas Observatory in Chile's Atacama Desert.

Colleagues are often surprised when I tell them this story because some of them tend to think of Paul as a bit gruff. When I asked why he bothered taking a theoretical physicist on a trip, and a talkative one like me who almost certainly got in the way, he told me that after writing the Press-Schechter paper for which he is famous, he couldn't get a job as a theorist, and he didn't want me to get stuck trying to do theory with no options. I probably smiled graciously while thinking, "OK, dude." And indeed, I was right that Paul would not convert me to an observer (although I have since slipped into doing some X-ray and gamma-ray astronomy), but I cherish, immensely, the great care he showed for my future in the field, and well-being in general. And though I stuck to theoretical physics, a conversation with Paul under the stars in the Atacama Desert changed my life.

The most important part of observing runs occurs at night when we capture images of the sky, with lots of sleeping in rooms with black-out curtains for much of the day. Paul made a point of showing me what was under the hood of optical astronomy. During the day, he spent a couple hours taking me through the structural guts of the twin telescopes at Magellan, explaining design choices and why he had helped push things in one direction or another. Most of what he said went right over my head, but it was there that I got an up close and personal view of the incredible scale of engineering required to build an observatory. In meeting the Chilean technicians who were there, I also realized what kind of partnership was required between the astronomers who flew in sometimes just for a couple of nights and the people who were rooted in the local community and stayed close to the land where the observatories are built.

One evening, Paul took me all over the mountain we were on so I could see some of the older observatories and meet some of the people working there. Magellan was the most modern, state-of-the-art facility there, so visiting the older telescopes felt like time traveling. But the other facilities, though smaller and with less modern-looking operating rooms, were nonetheless still operational, still being visited by astronomers who made incredible use of them. In my experience, no telescope is wasted on astronomers, although sometimes the land beneath it, and its stories, are lost on too many of us. At the end of the tour, Paul took me outside one of the facilities and for the first time—in the middle of the night, the farthest I had ever been from a city, and with some of the best seeing, meaning the least atmospheric interference on the planet—I saw the sky. Only once before had I seen the Milky Way, also known as "the rest of our galaxy"; the strip of stars runs across the sky and is comprised of something on the order of one hundred billion stars. But never had I seen it like this. To this day, I struggle to find the words to describe what I witnessed. As someone who doesn't really believe in the supernatural, I started to understand why people would. How could the universe just be—and be that majestic? As I stood there looking at more white dots than I could count and the white haze that inspired the Milky Way's name—a combination of stars that are too small for our eyes to resolve and illuminated gases—I felt filled with wonder, but also intense grief.

Growing up next to a freeway in Los Angeles, I thought it was pretty cool when I could see even ten objects in the sky. I got used to mostly seeing the moon when I looked up at night; in the mornings, when we were dashing to catch the school bus, sometimes I could see the planet Venus too. In the end, I made it through two degrees in astronomy, a PhD in physics, and a year at NASA before I got a sense of what the night sky must have looked like to my ancestors. They had seen the Northern Hemisphere's answer to *this* magnificent night sky. In that moment at the Magellan Telescopes, I knew my ancestors in a way I had never known them before, and I realized that many Black children were being denied a birthright, a human right, to know the night sky. And it was a reminder that for those of us who are descendants of slaves, which is to say most Black people in the Americas, another piece of our birthright that we have been denied is knowing how our ancestors understood and related to the night sky, away from the white gaze. While it is true that we can learn about the cosmologies of Africans, and we can also develop our own culturally distinct relationships with a variety of cosmologies, for us this will always happen, as literary and Black studies theorist Christina Sharpe puts it, in the wake of slavery, an institution that functioned in part by severing us from our Indigenous ways of knowing. The grief associated with this is permanent.

This grief has helped me understand why those who can still identify their home Indigenous communities often fight so hard to preserve not just their political autonomy but also their cultural autonomy. This means it's important to point out that I had this incredibly meaningful revelation on land that I had access to through a narrative that begins with violent colonialism. The irony is not lost on me that I experienced this realization in the Atacama Desert. Telescopes owned and primarily facilitated by academic institutions from rich, Northern Hemisphere nations dot the mountains there. The "seeing" is incredible—even as climate change shifts our environment, the atmosphere is more predictable, the conditions are dry—and it's perfect for telescopes, which have been popping up since the days when Chile was ruled by a brutal dictator who committed genocidal acts against the Indigenous peoples who had already survived the onslaught of European colonialism. Among those Indigenous communities were the Atacameños (*Likanantaí* in their own language), who to this day continue to dispute the distribution of land between them and the settler state of Chile. But, during the visit, I didn't think about those things. I had been told that there were no Indigenous people left in Chile, and I believed the story. I allowed myself to absorb a simple answer to a question that could never have a simple answer.

Just a few years after my visit to Las Campanas, I found myself in conflict with my professional community over the building of a telescope. In fact, many astronomers believed I was on the wrong end of what they understood as a fight between science and religion. That too was a simple and false story. That year, 2014, I learned that I was part of a professional community that was willing to inflict violence on Indigenous people just to build a telescope. Many of my colleagues, it turned out, didn't understand or care to understand what it feels like to be robbed of your identity and your home. This caused me to have an intense crisis about my identity as a scientist. I found myself asking, *What is the point of science if I have suffered so much for it? What is the point if it makes others suffer unnecessarily too?* It was not astronomers who provided a way out of the crisis for me, but rather the people they were demonizing. It was kanaka ʻōiwi (Native Hawaiian) cultural knowledge holders who—through challenging my status quo—forced me to look at science critically and imagine alternative ways of locating myself in science and science in the world.

The story of how begins long before, in 2001. I was at my desk in Adams House, my residence at Harvard College, checking my Yahoo! News page, when the email from Josh arrived. In an act of thoughtful generosity, Josh, a physics major a year ahead of me in school, was offering me a job as a telescope technician at a new observatory on Maunakea. Yes, *that* Maunakea. The position would pay around $50,000 per year, which was about twice what my financial aid forms indicated my mom had been raising me on before I went to college. My mind raced with possibilities. I gleefully assumed that I could pay off my student loans before I even graduated. And for once, I could tell my grandmother in Brooklyn about actually succeeding at something in college. I might actually pick up the phone, instead of shamefully avoiding calling her on all days except her birthday, anxious about my failure to achieve anything that she could gossip about to her friends.

This is my ticket, I thought. With one year off from school and experience at a major astronomy facility, I was guaranteed admission into a good PhD program. My mediocre grades wouldn't matter if I had letters of recommendation saying I had excelled during a year of substantive, full-time work in astronomy. With this one job, I could erase the humiliating reminder that I had been unprepared for Harvard. Rather than foster my aspirations to make a significant contribution to particle physics and cosmology, Harvard had taught me to see myself as a working-class kid from an overpopulated, under-resourced school district who could never win at an upper- (middle-) class man's game. I had spent the previous two years in a state of shock, increasingly traumatized by my apparent inability to negotiate power dynamics that seemed crushing and difficult to understand. With this one email, Josh had offered me the key to change it all.

I immediately started writing an email to my mom, so excited that I could finally make her proud. My mom, widely regarded as a brilliant strategist and advocate for grassroots women's organizations, had mostly raised me alone. I had a father and stepmother in the Washington, D.C., area, but they were mostly part of my life during the summers and briefly in the winters. When my mom realized that my final two years of high school were going to cost more money than she had as a low-income parent, she didn't tell me that I had to drop out of my nice magnet school to attend one that would allow me to work an afternoon job. Instead, my mom, a woman who knew her way around the United Nations headquarters in New York City, started working nights as a secretary. With this job and the doors it would open for me, I would be able to tell her that her sacrifices for me were worth it.

Midway through the email, I paused: I wanted to tell her a little about what the offer entailed, so I typed "astronomy Maunakea" into my Yahoo! Search box, excited to gather more information about my glamorous future. I'd been to Hawai'i before: a dear friend's parents had taken me to Maui the summer after we graduated from high school, and I was imagining beaches, amazing tans, and an opportunity to start over, not just intellectually, but also socially. Hawaiʻi would be the perfect getaway from the unwelcoming world of the Ivy League.

One of the first links to pop up was about protests on Maunakea. Apparently Native Hawaiians opposed to the building of the telescope were picketing daily. I read the piece with a sinking heart; I came from a family that walked picket lines, not one that crossed them. My mother was a grassroots organizer who would later be recognized for her effort to bring the serial killer known as the Grim Sleeper to justice. My father was a gas man turned union man, and before he moved away, I spent many weekends with him and my stepmother, a labor lawyer, walking picket lines filled with striking Hyatt Hotels and Eastern Airlines workers. There was absolutely no way I could cross a picket line, especially not one organized by Indigenous people. In the summers, my mother and I sometimes made the trek to South Dakota as guests of Lakota friends who hosted us during the Sun Dance. I knew honoring a people's relationship to their land was paramount, certainly more important than any curiosity or non-emergent financial needs I might have. Plus, just months earlier, I had been one of the members of the Harvard Living Wage Campaign, where I'd helped to organize a weeks-long sit-in and tent city in Harvard Yard that had made national news and even elicited a visit from John Sweeney, the head of the AFL-CIO. I had been furious with students who critiqued our cause—effectively crossing the picket line. I couldn't become one of them.

In fact, one of the reasons I had struggled so much at Harvard was that I had set myself apart from my classmates by being infuriatingly insistent that taking action was not just an extracurricular activity but also a moral duty for anyone with keys to the Harvard kingdom. After the sit-in, having barely passed my physics and math classes, I promised myself that I would make more room in my life for my dreams of being a physicist. But not like this.

So it was with the heaviest heart that I wrote to Josh that it would not be possible for me to take up the job at Maunakea because I could not cross a picket line. Josh quickly wrote back, saying that the situation was complicated and that he'd love to discuss it on the phone. But, as I had done many times before in my Harvard career, I girded myself against being talked out of abandoning a core principle. My response to him was terse, hiding the intense grief I felt at a missed opportunity: a picket line was nonnegotiable, thanks for asking.

As hard as that was, I never regretted my choice, but I also never took the time to investigate why there had been a picket line, perhaps to protect myself from dealing with the lingering bitterness I felt. Little did I know that thirteen years later, I was going to learn more about what the protesters—or protectors, as I now understand them—were doing on that mountaintop, more than I ever thought I'd know about anything besides physics. By the time I graduated from Harvard with a primary concentration (major) in physics and a secondary in astronomy and astrophysics, I had taken thirty-three courses, with twenty-one in physics, astronomy, math, and applied math, taking coursework outside my core area only when it was required or about ancient Egypt or Jane Austen. I wanted to be what I understood a theoretical cosmologist to be: a person who used math to understand where everything in the universe came from, starting with the origins of spacetime and the particles therein. I wanted to pick up where Einstein left off: asking questions like where did a curving and expanding spacetime even come from, why was the expansion accelerating, and why was it full of particles like electrons and quarks? It was never really my plan to become a navigator of late-time, earthbound phenomenology, otherwise known as scientist behavior toward those they considered to be nonscientists.

In December 2014, eleven years after I graduated from Harvard, a news item from a local broadcast channel based in Hawai'i was posted in the Physics and Astronomy Equity and Inclusion

Facebook group. A group of Native Hawaiians had been violently arrested at the building site of the new Thirty Meter Telescope (TMT) on Maunakea. I was thirty-two, and a postdoctoral fellow at MIT. I knew what was at stake scientifically. This was to be the biggest telescope ever built on Earth, located in the best possible place to see the night sky clearly. Maunakea is an unparalleled resource for those who wish to gather photons—particles of light—that will help them tell our cosmological story. My expertise taught me that the TMT was the next logical step in our grand quest to understand the universe, but I had not forgotten eighteen-year-old me and still shared her values: I understood that something deep was at stake socially and politically and that it was my task to find out what it was. Doing so would change my life forever.

One evening, I noticed that a Lakota graduate student in astronomy was tweeting about how the Facebook group conversation about Maunakea made her feel devalued and disrespected as an Indigenous person. Because she was someone I cared about, I felt it was important for her to see someone even a little more senior to her showing solidarity. So I wrote some tweets. It snowballed from there. I was contacted by the press, which forced me to think a lot about my own Caribbean roots, and how I would feel if white scientists showed up in Barbados and insisted that a piece of land that was sacred to us had to be used for their own purposes.

After a story started circulating about how the last King of Hawaiʻi, Kalākaua, had wanted telescopes on Maunakea (which, I later learned, is not really what happened), I realized I needed to read more about the history of astronomy in Hawaiʻi. I learned that the first telescopes were built on Maunakea at a time when Native Hawaiians had no say in how their land was used, and opposition to the facilities existed right from the start. I remembered a conversation I'd had with a taxi driver, a man who looked to be, like me, in his early twenties, while I was on an astronomy-related visit to Maui. I had asked him if he had ever been to "the mainland." Evidently irritated, he responded that he would never visit the United States because he was a Hawaiian. I understood well enough what he meant—and respected why he felt that way—but didn't think about it too much more because I was dealing with my own struggles as a Black woman, and because I didn't have to. But his comments were my first hint about what was happening. My second was an extremely condescending event during that same grad school visit, where astronomers were lauded for helping employ local Native Hawaiians as technicians. Why not employ them as astronomers, I had wondered?

As I learned more about Native Hawaiian narratives around what astronomy meant, I began to question the stories I had been told about why and how we do astronomy. The Air Force telescope on Haleakalā that I had visited as a grad student (with armed military police following our group around the facility) was funded not because we wanted to understand the universe, but because of the military. Twelve years of higher education had taught me not to ask out loud uncomfortable questions about the intersection of morality and my corner of science because scientists rarely responded well. Plus I felt helpless to change anything. But now I was in a position to say something and realized I didn't know enough to say anything in great detail. I began ordering history of science books, spending my weekends reading rather than relaxing or socializing. I learned that science history, as it had been taught to me by physicists and their textbooks, was at best only superficially accurate, but often wrong. At worst, it obscured a longstanding history of racism and colonialism in science. I found myself in a crisis from which it would take years to recover. My once firm faith that the real purpose for basic science was following our curiosity began to elude me. This was not just about the military, I realized, but also about colonialism and intellectual work as capitalist enterprise.

There is a strange contradiction among scientists: science is supposedly about asking questions, except about scientists and how science is done. Although research in the philosophy, history, sociology, and anthropology of science has gone on in earnest for decades, very few if any of the results are integrated into science education and therefore science practice. Scientists may not see how a subject like sociology or anthropology applies to their technical work, but science doesn't happen in a vacuum apart from society. Indeed, scientists are acting *unscientifically* when they do not acknowledge the history, philosophy, and sociology of their fields, which would help them understand how scientific research results have been used, abused, and imposed on people who were perceived as subhuman and unimportant. Instead, scientists demand that people accept their results without question.

The reality is that one could earn bachelor's, master's, and doctoral degrees in a scientific field and never take even one course that asks the student to evaluate the very process by which they produce knowledge. Sharon Traweek's 1992 comparative anthropology of particle physics experiments in the United States and Japan, *Beamtimes and Lifetimes*, provides an example of how research in science, technology, and society studies can have implications for how people do science and make science policy. Notably one of Traweek's findings was that women were excluded from physics both in the United States and in Japan, but because of completely opposing stereotypes: in one country women are considered to be too group-oriented and in the other, too independent. That these nationally based stereotypes are in conflict should be a red flag; maybe gender essentialism—the idea that genders and sexes have fixed, universal characteristics—is actually nonsense. The fact that women end up being seen as less valuable no matter what the stereotypes are suggests that the problem isn't our behavior, it's the fact that we're not cis men. And just like that, one can take an anthropological study of particle experimentalists in two countries and use it to change how scientific research in physics is done. Instead, our educational system produces scientists whose toolbox of technical skills does not include applying a scientific mindset to their work, their fields, or their very existence. It's also impossible for scientists to adequately adjust for the impact of their biases if they don't admit to having any.

In other words, many scientists tend to treat science and themselves like an infallible religious authority: the techniques and mild amounts of history we are taught in class and lab are like a religious text with only one interpretation. As a Jew, I laugh a little about this phenomenon, because our sacred liturgical task is to reread the entire Torah every year (and every Rosh Hashanah—I promise that this is the year I will actually do the whole thing) and then think about all the different interpretations we can glean from it. In a lifetime, if you actually get down to reading, you will think of many. As a scientist, I worry about the fact that our community doesn't do this enough with science.

When I took a course in waves and optics as a college sophomore, for instance, I never questioned why Snell's Law was named after Dutch astronomer Willebrord Snellius. I assumed it was because in the seventeenth century Snell was the first person to discover it. I didn't think to question it. So I was shocked years later when I found out that he wasn't, and that in fact the first person to find the law wasn't a European. My entire education had taught me that all meaningful science had come from Europeans and their descendants, when all along Snell's Law, sometimes called the Snell-Descartes Law, was in fact first discovered in the tenth century by Persian scientist Ibn Sahl. Using our own naming conventions, the law should be called Ibn Sahl's Law.

The implications should be obvious: how we talk about Snell's Law changes how students see science, and in this particular case, how people of Iranian or other Middle Eastern heritages might imagine themselves in the historical narrative. But like disgraced and fired Google employee James Damore, infamous for his memo to his former Google coworkers, there are many in the STEM industries who view concern over equality and inclusion in the discourse of people traditionally seen as weak, such as women, as political correctness. This strand of thinking views the search for narratives that highlight non-European participation in the development of scientific ideas as desperate opposition to meritocracy. As a Black woman I'm primed socially to hear these attacks and worry, because my whole life I've heard the message from a white, male-dominated society that trying to see myself in the halls of history is a character failing. But as a scientist, and by that I mean someone who seeks to look past biases to produce convincing and accurate models of how the universe works, the "political correctness alert" reaction is worrisome in its anti-empirical character. Isn't it just true—a matter of acknowledged fact—that Ibn Sahl discovered the law first? Isn't naming it after Snell without ever acknowledging this exemplary of an anti-meritocratic bias in science?

I joke to my friends that I fight scientists with science. It is a strange place to find oneself, fighting scientists about whether it's OK to ask questions about science or even to simply offer corrections about how science discusses its own history. Strange in part because the scientific community already accepts that at least some corrections are acceptable—in time. We have come to recognize, and accept as an allowable course correction, that Einstein's theory of gravity is more complete than Newton's. On the other hand, certain ideas that should be dismissed have resisted the same course correction. For example, even though most scientists believe eugenics is wrong, biologist James Watson still gets speaking invitations where he shares his eugenicist ideas. Likewise, research suggests that a significant number of American medical students still believe that Black people don't feel as much pain as white people do. These beliefs are irreconcilable with empirical facts. Yet, the hateful and angry emails I received after I published an essay in *Slate* about scientific racism suggests that it is still not entirely acceptable to ask the question, "What is the relationship between biology and racism?" For too many within and adjacent to the scientific community, the answer is foregone, that there is none, and the punishment for asking includes having my entire department receive an email explaining the intellectual inferiority of Black people.

A deeper interrogation is necessary: what is science and is it always independent of what we want to believe? Or does our desire to hold on to certain truths shape our science? Biological anthropologist Jonathan Marks writes in *Why I Am Not A Scientist: Anthropology and Modern Knowledge* that "science is the production of convincing knowledge in modern society." He goes on to define "modern society" as ideas that arose from Europe during the Enlightenment. But what about science outside of a European context? As Laurelyn Whitt describes in the introduction to *Science, Colonialism, and Indigenous Peoples: The Cultural Politics of Law and Knowledge*, constructing a definition of science that recognizes "science-like" activity and ideas outside of a European context without colonially forcing a foreign concept on Indigenous peoples is a challenge. Whitt argues, and I agree, for a move "to speak of a knowledge system . . . to abandon the idea that a single epistemology is universally shared by, or applicable to, all humans insofar as they are human."

But the mathematical laws of physics are universal, one might say, and those constitute a single knowledge framework, darn it. Perhaps, but it's possible for two geographically distinct communities (i.e., Europeans and the Palikur of the Amazon) to use completely different geometries to develop self-consistent descriptions of the same changing night sky, as anthropologist Lesley J. F. Green describes in her article "Challenging Epistemologies: Exploring Knowledge Practices in Palikur Astronomy." While the Palikur do not call their astronomical system a *science*, it nonetheless allows them to predict when one of the five rainy seasons of their environment will be arriving, to the day. I wish we could get such good weather forecasts in Boston.

THE REALITY IS THAT *DEFINING* SCIENCE IS NEARLY AS MESSY as doing it. That shouldn't be seen as a threat to science, but rather part of the challenge and the joy of the process. There is so much to learn, and the world of science, technology, and society studies has so much to offer students and practitioners of science, if we let it. But we also have to understand what responsibility we have when doing science and using the technologies created by science, especially as we begin to think about the possibility of sending humans beyond the moon to other heavenly bodies like Mars. In 2018, Spanish artist Santiago Sierra told the *Guardian* (UK), "Planting a national flag in a hitherto unvisited place has never been an innocent gesture. This is how colonial processes always begin."

I know that when colonial processes begin, no one is thinking of what futures are precluded by them. Here, then, is the list of anxieties that I experience when I think about humans arriving on another planet in corporeal form:

- Could mining on the moon forever alter how future generations see this majestic satellite of Earth?
- Who will profit and who will perish in the expedition?
- Will such missions ultimately exacerbate inequalities that exist on Earth? Will missions to Mars make life on Earth worse?
- How can we have these conversations with a language that refuses to untwine itself from the lingua franca of settler colonialism: "discovery," "exploration," "settlement," and "colonizing"?
- What are possible future generations on Mars? By this I do not necessarily mean human or even Earth-originating life. Who and what do we eclipse by choosing to lay claim to land?

To think about these questions, some people begin by reading revered texts by and about Black revolutionary thinkers like Claudia Jones, C.L.R. James, Frantz Fanon, and Aimé Césaire. I think about them first as an annual Star Trek Convention gold ticket-holding, seen-every-episode-of-every-series fan. While I think the franchise is imperfect, on the whole I have loved its expansive vision not just of humanity but of humanoidity—all of us humanoids learning how to share not just a planet but a galaxy. As a physicist, I spend some time at the conventions explaining to fellow fans that warp speed travel—travel at the speed of light—is probably not going to happen. And this means, the key generative moment for first contact in the *Star Trek* universe—humans reaching warp one and catching the attention of the Vulcans—isn't possible. It's likely possible that even if the Vulcans did exist and had better space travel technology than we do, they still can't travel at warp either. More likely, they've figured out how to live multigenerationally on a ship that could take centuries to get anywhere. Then again, physicists have often been wrong in the past about what is possible, so maybe we will be wrong again.

While planning remarks about human travel to Mars for a September 2018 panel at the Library of Congress, I reached out to Connor Trinneer, who played the engineer Trip Tucker on *Star Trek: Enterprise*. I was especially keen to hear from Connor because *Enterprise* is the series that deals with Earth's first forays far from the solar system and how humans go on to form the storied Federation with the Vulcans. What if a planet didn't have animals but the air was filled with psychedelic pollen? What if having sex with an alien woman got a human cissex man pregnant? These were just some of the scenarios he had to get into character for so that Trip could grapple with them—and all of them had to do with the ethics of making first contact with alien worlds. I expected Connor to mention something about the Prime Directive—the Vulcan rule adapted by the Federation that teaches us not to interfere with the internal development of other species. But instead he said something that perhaps reflects his childhood in a community that had a strong presence of both Indigenous and white people: he always thought that if we did meet another species, it would be because they wanted something from us, and that could get uncomfortable pretty quickly. I hadn't thought about that, even though I guess it's the storyline for almost every alien film.

Much has been written by Indigenous scholars both here in the Americas and elsewhere about how this style of storytelling reflects a kind of white anxiety that one day they might become victims of the kind of colonial apocalypse they visited on Indigenous people in Africa, the Americas, the Pacific Islands, Australia, New Zealand, and Asia—even sometimes on the European peninsula of Asia. As Black science fiction and fantasy writer Nalo Hopkinson points out in her introduction to *So Long Been Dreaming: Postcolonial Science Fiction and Fantasy*, "one of the most familiar memes of science fiction is that of going to foreign countries and colonizing the natives . . . for many of us, that's not a thrilling adventure story; it's nonfiction, and we are on the wrong side of the strange-looking ship that appears out of nowhere."

But what struck me about Connor's response to my question was that we could be the ones who want something. We could be the ones who want something from the moon and Mars (and as of September 2020, Venus)—perhaps to buy time here on Earth by extracting resources that we refuse to learn how to conserve? In wanting something, do we eclipse futures that we are not competent to imagine? Do we violate the Prime Directive before there is a civilization to disrupt? Do we relegate certain ideas to the past, therefore foreclosing on their potential to impact the future? Much as my research is about curiosity and imagination, my day-to-day realities as a scientist are also deeply tied to science as a tool that enhances colonialism. Astronomers were once being funded to watch an eclipse from Haiti in order to better measure distances, ensuring that slaves and goods would move faster across the Atlantic. Today they are funded to help develop adaptive optics, a military technology that makes our beautiful pictures clearer, whether they are spy images or galaxy images.

It is in this context that I ask a question we should always be asking when our science and technology push us in radically new directions: why now? It should not be surprising that asking "why" has led me as a scientist to other questions. Do we have the right stuff to connect with other lands when the most powerful among us refuse to acknowledge our relationship to the lands we know? Is this the right time in human history for this new way of being in the solar system? What meaning does a mission to Mars have for making Black Lives Matter? For protecting Indigenous women and girls from kidnapping and murder? Will it keep our water safe? The promise of technological advancement that will improve lives is tantalizing—but then what of Ferguson? Is a pipeline through Standing Rock progress for the people of that land? Is the fourteenth telescope on Maunakea, one that will be more permanently destructive to the mountain than any before it, progress for the people of that land? Is the global warming we have achieved through technology "progress"? Is our relationship to progress catastrophic?

This brings me back to Maunakea. The colonial legacy of astronomy has taken us from Europe to Haiti and now to Hawai'i. Hawai'i is the flash point for this conversation now because of the visible violent threat of force against the Mauna's protectors; in 2015 and then 2019, I repeatedly found myself begging astronomers to oppose the use of the police and military to enforce their colonial right to build on sacred land. But the story goes beyond the Thirty Meter Telescope. It requires scientists to learn and understand Hawaiian history. And we must understand that Hawai'i is not the first or only place where astronomers used and benefited from colonialism.

I BELIEVE SCIENCE DOES NOT NEED TO BE INEXTRICABLY TIED to commodification and colonialism. Euro-American imperialism and colonialism has had its (often unfortunate) moment with science, and it's time for the rest of us to reclaim our heritage for the sake of ourselves and all the generations after us. That doesn't mean changing ourselves so we can be included in science—it means changing institutionalized science, so that our presence is natural and our cultures are respected. Yet even as we begin to think about the importance of making these changes, we are also confronted with the way dominant forces tend to co-opt the language we develop to talk about our experiences with oppression. "Decolonization Is Not a Metaphor" by Eve Tuck (Unangax̂) and K. Wayne Yang is one of the most thoughtful discussions I've seen anywhere on the increasing popularity of the term "decolonization" in twenty-first-century American diversity discourse. In their article, Tuck and Yang highlight the very real threat posed by co-optation of "social justice" language by a more establishment-oriented liberalism. By not carefully considering the word's meaning, for example, ignoring the full scope of colonialism, adopters of the term can end up stripping it of its value.

Importantly, colonialism isn't just about land—it's also about culture and ideas. The residential schools in the United States and Canada that kidnapped First Nations, Métis, and Inuit children and forcibly stripped them of their culture, language, and even hair while making many victims of violence (including sexual abuse and rape) were furthering a colonial project of violently attacking what made Indigenous people who they were. The denial of language and history is a fundamental feature of the colonial project—in part designed to minimize the possibility that people will feel empowered to push back against and end colonialism. This is part of the power behind the name of the #IdleNoMore protest movement that arose in December 2012 among Indigenous people in Canada. Colonialism was meant to strip people of their identities and in the process make us idle in the face of oppression. In 2012, when the Canadian government again attacked Indigenous ways of life, people rose up and said, "We will be idle no more." Similarly the #WeAreMaunaKea movement to stop the building of the TMT has been born out of an anti-colonial cultural revival: many of the kānaka ʻōiwi who are leading the charge are from a generation who have benefited from their elders' fight to bring Hawaiian language and culture back into the schools—and to build schools that were specifically designed to do this work. Many of the Mauna's protectors are products of these schools.

In many ways, these cries for justice parallel Black cries for freedom, our fierce protectiveness of African American vernaculars and Black diaspora creoles, and the fight for Black studies. We too oppose the colonization and erasure of our identities. Tuck and Yang are part of a long line of thinkers who sought to explain what it would mean to eschew inclusion in the colonial enterprise and instead strip colonialism of its power. Black revolutionaries like Frances Ellen Watkins Harper, Ida B. Wells-Barnett, and C.L.R. James all expressed anti-colonial sentiments, even if they didn't all use the language of colonialism. Black people in the Americas are too often forgotten and even actively ignored in discussions about colonialism, but the colonization of our bodies and the severing of our ties to our culture is part of how land theft was enforced, maintained, and made profitable. If we are to talk about decolonization, we cannot be piecemeal about it. When we talk about decolonizing science (or anything, really), upending settler colonialism everywhere is central to that project. I cannot separate telling the truth about the history of Native Hawaiian astronomy from the fight to stop settler scientists from using the history of Native Hawaiian astronomy to excuse colonizing Hawaiian lands. I cannot separate wanting to tell the truth about Islamic contributions to physics from also having complex conversations about Indigenous sovereignty and what that means for me as a descendant of one of the kidnapped ones.

I came to the discussion about decolonizing science due to the controversy about the uses and abuses of Maunakea. To this day, I have never been up to the Mauna. I did not go before the publication of this book. I did not make a point of going because I was not needed there, and it is not my ancestral land. I was not needed on Haleakalā the time I visited the telescope there, like an earthbound space tourist. I went there and did not know that it was contested land, did not think to ask about it. I am learning to think better. During the 2015 efforts to protect Maunakea, I saw in photos that some kānaka ʻōiwi carried signs saying, "Pono science is possible," a phrase that was first articulated by Kalei Lum-Ho. *Pono*, as I have been taught by my friend Professor Bryan Kamaoli Kuwada, is tied to the practice of *Hoʻoponopono*. I understand Hoʻoponopono as activity that brings about healing and balance, with pono the outcome of that. Hoʻoponopono is about making things right. Pono has many different, imperfect English translations, including righteousness, virtue, goodness, and hope. As kanaka ʻōiwi biocultural scientist Katie Kamelamela put it on Twitter on July 14, 2019, "Astronomers have no ethical code because they 'don't work with people' . . . this needs to change."

I appreciate Dr. Kamelamela's advice, and I regret the conditions under which she felt compelled to offer it. I also know we cannot turn back the clock. In his article "Rhetorical Sovereignty: What Do American Indians Want from Writing?" Scott Richard Lyons (Ojibwe/Mdewakanton Dakota) articulates that turning back the clock isn't even the point, when he says, "the pursuit of sovereignty is an attempt to revive not our past, but our possibilities." The creolization of communities that has been produced through settler colonialism, globalization, and imperialism can never be undone. What *can* change, however, is our relationship to the current power arrangement and our relationships with each other. In this context, the question of what it means to do science is to ask about science's relationship to power. Is power the point of science? Power over what? Power over whom? Thanks to the *kiaʻi*, Maunakea's protectors, I now understand that I must demand of myself an empirical practice, a science, that is not organized around power. Instead, I work daily to understand myself as a quark assembly of supernova remnants on a journey to know and honor all our galactic relations.

THIRTEEN

COSMOLOGICAL DREAMS UNDER TOTALITARIANISM

> He has now reached the stage where he sees the problem philosophically, as a problem of world civilization. How to reconcile the undoubted advantages of an industrial civilization with what that very civilization is doing to him as a human being.
>
> —C. L. R. James, *Mariners, Renegades, and Castaways: The Story of Herman Melville and the World We Live In*

SCIENCE AND TOTALITARIANISM—THE UNDEMOCRATIC, FAScist exercise of total authoritarian power over communities of people—have typically had a pretty cozy relationship. This is difficult for many scientists to accept. Certainly, when astronomer Sarah Tuttle, biophysicist Joe Osmundson, and I published a statement—"We Are the Scientists Against a Fascist Government"—a month after Donald Trump was inaugurated in 2017, I felt that we were years overdue, but there were people who suggested we were overreacting. The March for Science that happened a few months later was stridently against openly admitting that Trump was probably a fascist, insisting that it was nonpartisan and for a while trying to convince everyone that a March for Science under Trump was somehow not political.

I understand, deeply, the impulse to separate science from human relations. I'm pretty sure part of what attracted me to science as a ten-year-old is that in addition to the promise of a career, there was also the promise of something that operated beyond the grassroots concerns of my activist parents and the people whose lives they were trying to save. I wanted my life to be spent on the dramatic glamor of the cosmos, not the mundane horrors of how human beings treated each other. I wanted science to be like the film *Real Genius*—full of practical jokes (an old Caltech tradition), sexy nerds, and ultimately, young scientists who make the noble choice in the face of financial coercion from the military-industrial complex. But as Bob Dylan once said, "We live in a political world." When I'm not feeling cynical, I smile when I recall how adorably naive I was. I thought I was working to change the world, when instead I had consumed incredibly effective intellectual propaganda.

So I understand why other scientists don't want to be confronted with the fact that science is inextricably tied to everyday, human, social phenomena. These feelings of refusal are heightened for people who have significant socioeconomic ease and political capital relative to others: no one wants to hear that their noble calling is actually deeply entangled with white supremacist ableist heterocispatriarchy. But it is.

Far too many white scientists whom I know—including people I respect, would even call friends, and certainly am happy to work with—seem really surprised by the horrors of the Trump era, which will almost certainly continue even after he is gone since some of it predates his rise to the presidency. I find myself wanting to tell them over and over again: welcome to the show. This is terrible. But it's been terrible for a while. It's clear that until Trump was elected, large swathes of scientists never felt like things were so bad that they needed to march in the streets. And when they did finally organize in large numbers, as scientists, it wasn't for Black lives or Indigenous sovereignty and water rights; it was for themselves, for a science that has traditionally marginalized Black and Indigenous people.

Even in the face of totalitarian catastrophe, scientists were more concerned with the appearance of civil, nonpartisan respectability than with confronting the fact that "civility" often meant death for people with connections to the Global South, including poor communities of color in the US. Of course, many scientists studying climate change are trying to save the planet, and that involves all of us. But large-scale organizing outside of the Black community didn't happen for Black lives until 2020, even though we are among those who are most at risk. Meanwhile, I grew up in East LA, where ICE regularly kidnaps people now and where ICE's predecessors were always a threat. I went home every day to a neighborhood where the police chased people through my backyard and there were shootings and people died, although luckily for me never anyone I knew. But death was normal. Police violence was how things were. In other words, I never thought things were great for Black people, Indigenous people, and Brown Latinx people because I'm a Black kid from Mexican and Central American East LA who grew up spending time with Lakota folks in Los Angeles and South Dakota. I have always known better. But it's a delusion that a lot of my white peers seem to have experienced right up until November 8, 2016.

To understand how this works, it is helpful to remember Noel Ignatiev* and Ted Allen's conception of "white-skin privilege," which they developed in the 1960s. These days "white privilege" is often packaged through a liberal framing: it can be thought of as having such easy access to one's rights that one never feels compelled to stop and consider that many others can't access those same rights so easily. This conception is a far cry from Ignatin and Allen's original Marxist analysis, which articulated how white-skin privilege was a tool of rich capitalists, a way to encourage

* Ignatiev was writing at the time as "Noel Ignatin."

white working-class people to side with rich white people. In other words, whiteness exists to serve capitalism by making pan-racial working-class solidarity difficult, and it harms not just people of color, but also working-class white people.

Years before Ignatiev and Allen advanced this anti-capitalist framing, W. E. B. Du Bois developed a similar line of thought, describing the "psychological wage" of racism in his book *Black Reconstruction in America*. He wrote: "It must be remembered that the white group of laborers, while they received a low wage, were compensated in part by a sort of public and psychological wage. They were given public deference and titles of courtesy because they were white. They were admitted freely with all classes of white people to public functions, public parks, and the best schools. The police were drawn from their ranks, and the courts, dependent upon their votes, treated them with such leniency as to encourage lawlessness. Their vote selected public officials, and while this had small effect upon the economic situation, it had great effect upon their personal treatment and the deference shown them."

Today whiteness, especially when paired with middle-class or higher socioeconomic status, also serves to make anti-fascist solidarity difficult. Because of white-skin privilege, many white people believe they have experienced America as a democracy, not a settler colonial state governed through racism. It is almost like living in two different realities, where white people were given sufficient amounts of freedom that they were willing to ignore the ways in which even many of them are not totally free and the ways people of color were even more restricted. Only during Donald Trump's blatantly racist, xenophobic, and misogynist rise, paired with his evident interest in antidemocratic, authoritarian approaches to governance of even white lives (and distaste for science), were people of white-skin privilege, including many scientists, confronted with experiencing diminished rights now long familiar to so many US residents and citizens of color. What is it like to be descended from a kidnapped people? What is it like to be descended from peoples who had their languages systematically and violently stripped from them, along with access—freedom—to know and live on their homelands? What is it like to be systematically denied access to health care, employment, or housing because it isn't in the interest of rich capitalists for you to have your fundamental needs fulfilled?

In the America where Donald Trump can become president, many Americans have "discovered" for the first time, as if in Christopher Columbus cosplay, that America's political foundation has a fascist, totalitarian pulse. In the months leading up to the 2016 election, many of us said, "Listen, can't you hear the telltale beat of racism?" But many Americans are so frightened of an honest confrontation with the present that they pointed to Barack Obama, as if he was singular proof that racism actually wasn't all that powerful. Many of us responded by pointing to the police shootings and literal lynchings of Black boys and girls and women and men and nonbinary people. We also pointed to the high number of deportations under President Barack Obama and his extensive use of executive power to murder people in the Middle East, Africa, and Asia with drones, paired with an unwillingness to use those same powers to save the lives of undocumented people. In my view, to be an American president is to helm a sprawling system that does not understand freedom even as it drones on and on about it.

In this context, what does it mean to speak of "American science" and "American scientists"? The United States began because European colonists (primarily men, but not just the men) weren't happy with the terms of engagement with their royal overlords, and they figured that Native people weren't worth caring about so they could just make a new country on their land, primarily using labor that was violently extracted from kidnapped Africans and their children and their children's children, the children who Black women were forced to produce. Slavery and rape go hand in hand. After that a lot of horrors happened, not just to Indigenous American folks and Black folks but also Brown Latinx, Asian, and Indigenous folks of other lands who came to these lands after the end of slavery. There was war, more colonialism that went as far as Hawai'i and the Marshall Islands and the Philippines, and imperialism that went as far as the southern tip of South America. The United States did business with Hitler until it had no choice but to go to war with him. We don't talk about how many Jews died because the United States turned them away. The Romani that were heavily discriminated against before Hitler's rise—and who were murdered by the Nazis—we don't really talk about at all.

In *Race and the Totalitarian Century: Geopolitics in the Black Literary Imagination*, Vaughn Rasberry makes the case that the racial legacy of Euro-American settlement and nation building in the United States is one of totalitarianism. This is what some white liberals finally got hip to in the days following the 2016 election: for people of color, this country has always been at least a little bit fascist. I remember that on election night I posted to Facebook something along the lines of, "We tried to warn you." A white astronomer I knew suggested to me that because I lived in Seattle, which was supposedly nice and liberal, I would be fine. He hadn't noticed that just in that last year, multiple Black families in the Seattle area had been targeted with notes telling them to leave their neighborhood, including in one case having "nigger" spray-painted on their minivan. During my first week of work as a postdoctoral researcher at the University of Washington, I sat down on the bus only to find the following written on the seat in front of me: "Fuck Black Life Matter. Fucking Niggers criminal."

The astronomer who thought everything would be fine and with whom I had previously had a collegial relationship, assumed that in the coming months no one was going to shoot their Sikh neighbor while he was in his driveway working on his car, which is what happened in a Seattle suburb not long after the election. He assumed we lived in a country that had never existed. In November 2016, Langston Hughes called out from his grave, "America never was America to me," and white liberals shouted back, "Make America Great Again! Impeach Donald Trump!" Meanwhile, the day Trump was inaugurated, a white supremacist shot a protester just a couple of buildings away from my office at the University of Washington.

The "discovery" for many self-declared liberals that America has fascist tendencies and that they, as white (mostly gentile) people, were not entirely shielded from this seems to have been quite the shock for many who believed in the American freedom mantra. After all, it was not their families, or families at all like theirs, who were experimented on during the Tuskegee Experiment. Their children had never been torn from them and sent to residential schools. Their children weren't disproportionately sent into foster care. Their mothers weren't disproportionately pathologized and criminalized for being poor. Their parents weren't kept from traveling to parts of Europe because of fears of "illegal immigration."

The United States is a system that begins with land dispossession and enslavement in service of one racialized group, at the expense of "the others." Via this system, the American state is able to exercise total control over the lives of people of color, especially those of us who are Black or Indigenous. Our opposition groups—Black Lives Matter, the American Indian Movement, the Black Panthers, The Young Lords, sovereign Indigenous nations—are primarily outlawed as terrorist organizations, or forced to operate under rules created by a state that is determined to strip its opposition of the power to effect change. Ironically, it is the US government that has, under COINTELPRO, unethically targeted these organizations with attempts to terrorize and divide its members and the communities that support them. In this system's sphere of influence, Black children cannot safely rest on the couch without being murdered in their sleep by police. Black children cannot go to the store and buy candy without being murdered on the street by vigilantes who are operating as part of a surveillance structure encouraged by the state. Black children cannot listen to music in a car without being murdered by vigilantes who believe the state gives them permission to shoot loud Black children. American white supremacy is a total authoritarian structure that shapes every aspect of Black lives.

We tried to tell white people about this. The ones who were listening were mostly the ones who thought it was a good idea. The others, the ones who benefited from it all but found the idea of it to be unsavory, buried their heads deep in the sand. And even after they discovered America's heart of darkness, they would still insist it was brand new and reversible. A recently introduced defect, rather than the way things have always been. They are in fact so frightened of an honest confrontation with history that modern race discourse has largely supplanted discussions of racism with analyses of "bias," which describes the same exact outcomes as "racism" but in a way that softens the language and ignores historical, structural context. It disconnects racism today from the genocide against Indigenous peoples, or the fact that Africans were kidnapped and forced into multigenerational slavery. This is of course pretty uncomfortable history to recall; frankly, there's something wrong with you/your worldview if it doesn't evoke feelings of discomfort.

WHETHER OR NOT SCIENTISTS ACKNOWLEDGE THIS HISTORY, it is a truth that American science has benefited from and been part of the process of colonialism and white supremacist development both here on this land and around the world. We'll get to the ways science has reinforced American society's imperialist projects later in the chapter. But one important way it happens is through its own intellectual colonialism, insisting that there is only one scientific vision: internalist science. Most of us are trained in this point of view but probably most of us haven't heard of it. Internalism insists that social phenomena don't influence science.

It turns out that this is not the only way to look at the world, as the Maunakea protectors have been trying to teach us. Scientists have responded with the full force of state violence behind them that scientific research and experimentation is about objectivity. But this objectivity they are talking about is a mythological concept that works as an effective tool of the colonial project. An externalist view, which holds that facts can be believable even if a thinker can't deduce those facts from internal logic and experience, would admit that the sociopolitical environment also impacts what constitutes science and what constitutes accepted scientific truth. Internalism, on the other hand, is the perspective that unless a thinker can determine whether something is true from their own internal sensory experience, they cannot conclude that something is true. This perspective, which is widely held both within and without the scientific community, has vast implications in a world where sensory experiences are necessarily diverse and where the distribution of those experiences tends to correlate heavily with ascribed identities such as one's race, gender identity, sex identity, and gender expression, as I described throughout Phase 3.

How science happens, and in particular scientists' commitment to internalism, can change. This would require confronting some ugly truths about how science includes and excludes people. It means that all scientists must ask themselves: what do we expect people to give up in order to succeed? What requirements about social background, clothing, hair, English dialect, neurotypicality, state of (dis)ability, and gender normativity do we need to excise from our community standards? What if scientific leadership took meaningful steps to eviscerate the powerful hold rape culture has on our community? How do we begin to repair the damage the dominant scientific community has done to marginalized communities and their relationship with the intellectual pursuit of understanding the world in a formal, mathematical way? How can we become more honest about the reality of subjectivity as a feature of the scientific community that cannot be eliminated? How can we promote the health and well-being of the most marginalized, for example, those who are both trans and disabled? If everyone was schizophrenic, the world would probably be built around the needs of people with schizophrenia, right? What are we failing to imagine when we refuse to understand that there is another, better way?

SCIENCE'S PROBLEMS ARE NOT LIMITED TO WHAT HAPPENS TO the people in it, however. There is also the difficult question of our relationship to the larger society that we are in. If our society is defined by white supremacist heterocispatriarchal capitalist values, science must contend with how it is shaped by those values. This was frightening to me when I was an undergraduate. I worried that the luxuries at Harvard would seduce me into forgetting that I rejected capitalism and militarism. In fact, I didn't consciously comprehend it in these terms at the time, but I was afraid of becoming J. Robert Oppenheimer.

Oppenheimer was born in the late nineteenth century to the "good" kind of Jewish immigrants—Germans who had made lots of money—and he grew up coddled, sent to private schools that prepared him for his time as a student at Harvard College, University of Cambridge (where he once told someone he tried to feed his mentor a poisoned apple), and the University of Göttingen. Oppenheimer went on to become an accomplished theoretical physicist, with an equation named after him that I use in my work on neutron stars: the Tolman-Oppenheimer-Volkoff equation. But he is most famous not for his science but rather for being the leader of the Manhattan Project in Los Alamos, New Mexico, where humanity's first nuclear weapons were developed. While hundreds of scientists contributed to the making of "The Bomb," Oppenheimer is the man who made sure their collaboration worked and was ultimately successful. Theories abound about why Oppenheimer committed to the project. I like theoretical physicist Isidor Rabi's proposal that Oppenheimer was obsessed with proving that he was an American because he could not accept himself as a Jew. At the same time, I was told by the late Rose Frisch, a scientist in her own right who was at Los Alamos because her husband David Frisch was one of the nuclear physicists working on it, that "Oppie and the others" hoped that it wouldn't work. As the people of Hiroshima and Nagasaki learned in August 1945, it did.

Oppenheimer came to regret ushering the world into the nuclear weapons era but also continued to remain deeply entangled with the American military-industrial complex, until, under the fascist leadership of FBI Director J. Edgar Hoover and Senator Joseph McCarthy, the establishment turned on him. In 1949, Oppenheimer was called before the House Un-American Activities Committee, where he confessed to past associations with the Communist Party and he snitched on several of his former students, more than one of whom suffered professional repercussions, including blacklisting. Oppenheimer was able to continue his life as a scientific leader for a few more years until his political rivals gained power, and he was brought before the Senate, which caught him in lies—because he had only half-snitched the first time he went before Congress.

Oppenheimer is a fascinating and terrifying figure. Like mine, his childhood was full of humanistic Jewish teachings, and his early childhood had similar influences as Mordecai Kaplan, the founder of the Reconstructionist Jewish tradition in which I practice. How did he end up going so astray? How did he cleave more to an institution than to his humanist values? I wonder if his transition to leading the production of the first nuclear weapons was facilitated by his scientific training. I wonder if, where others might have objected, he did not because he had grown up rich and sheltered in institutions that inculcated concessions to their authority. He had become white, and his whiteness gave him a comfortable proximity to power, a power granted by a system that he wanted to protect. A century after his birth, I see many of my peers struggling to disentangle themselves from structural whiteness, to see the system for what it is.

Being a Black woman and descendant of slaves has meant, maybe, that I couldn't fall for the trap so easily, although Black people sometimes think that they will find liberation in proximity to whiteness, and I have feared that I too would get caught up in this. But, in the end, I never came to share the belief that America was a sanctuary, and it's a good thing, because the United States has always been a totalitarian state for many people. Total authoritarian power has been exercised, unrelentingly, over the lives of Black and Indigenous people for centuries. And to my white friends and fam who are just realizing how bad things can get, I say again: welcome to the show. It's bad now, but the old way of doing things wasn't working for many people. And importantly, it is never going to get much better until there is significant structural change in the power dynamics that dominate North American society—and global society. Until there is a reckoning with the reality of the world in which science is done, only a small elite will be able to succeed economically. And only a few will be able to spend their days at colleges and universities, supposedly doing science.

Anyway, that's actually not what scientists do all day, especially at academic institutions. In addition to the service I discussed in "Wages for Scientific Housework," an inordinate amount of time in science is spent begging for money. At rich institutions, the faculty spend less of their time doing this, but still a significant fraction of "research" time is spent writing grant proposals. Success in science is so tied to money that some of the most prestigious prizes are not awarded because of recognition for contributions but effectively because someone wrote a very good grant application.

Importantly, areas like mine—often called "basic" science because of the lack of immediate social applications—have never received fiscal support just because humans are curious. The current funding structure for particle physics and cosmology can be traced to Oppenheimer's greatest regret. After the advent of nuclear weapons and a host of other technologies that assisted the US during World War II, the US government recognized the value in paying for nuclear physics and the subdisciplines it was spawning, namely particle physics. The problem is, public, civilian funding for particle physics peaked in 1968. By 1972, it was half that, and it has been in decline ever since. For particle theory, the decline has been particularly steep. A glancing look at budget appropriations makes clear that every year a little bit more of our budget is transferred to the "quantum information" column, meaning toward work on technologies that have obvious military and commercial value.

In the 2020s, we are facing the collapse of public funding for basic science overall and esoteric areas like particle theory in particular, the end of state subsidies for it through support for public universities (which are effectively being privatized now), and the rise of billionaires and near billionaires as major players. The lack of government funding for public universities and some areas of scientific research is in fact tied to the wealth of the ultrarich. They aren't properly taxed, and their donations to private institutions are tax deductible. This makes a difference for public universities in states that tend to be under-resourced, like my home institution of University of New Hampshire. Schools in New Hampshire (including, bizarrely, the Ivy League Dartmouth College) qualify for special funds because we are in an under-resourced state. The pot of money those funds are allocated from is simply smaller because rich people don't pay what I consider to be their fair share of taxes. It's easy to say that it's lovely they are giving it away via private foundations (and on more than one occasion this private money has supported my work, including research on racism in science), but ultimately that's an activity that sidesteps democratic governance. For example, while federal agencies are banned from taking our political views into account when making funding decisions, private foundations can do whatever they want. There isn't even a veneer of democratic accountability.

The relationship between science and exploitative profit is not new, of course. Although it is shaped in the twenty-first century by the decline of what was effectively a public-funding blip in the twentieth century, if we look back through the last five hundred years, we find plenty of examples of science being privately and publicly funded with profit as a goal. I've mentioned the seventeenth-century expedition to Haiti, supported by the Paris Observatory under the directorship of Giovanni Cassini, for whom a NASA space probe is named. Cassini also managed an astronomical expedition to French Guiana. His son Jacques was involved in a similar expedition to Martinique. The European astronomers who tagged along on expeditions to different colonies did so partly on the argument that the observations they would do along the way would make future trips more efficient. At the time, governments were particularly interested in how to quickly move around goods—including kidnapped Africans, some of whom were my ancestors. What we call "modern astronomy" has never been pure. It has always been tied to violent, colonial, and capitalist activity.

In fact, we believe in this pure ideal only because during the Cold War, scientists and government officials carefully crafted propaganda about the nature of physics in order to affect the opinions of both the public and lawmakers. As outlined in Audra Wolfe's book *Freedom's Laboratory: The Cold War Struggle for the Soul of Science* and Daniel S. Greenberg's *The Politics of Pure Science*, those seeking to prop up the newly dominant physics community promoted particle physics as pure, apolitical science, the kind that could only happen in a "free" and "democratic" society like the United States. This kind of messaging served the needs of scientists who wanted more funding allocated to their research, and it served the United States government by suggesting that an apolitical, "pure" science was only possible in capitalist, politically democratic America. This history points to an evident irony in the various claims made by leaders of the March for Science that it was apolitical and then when they dropped that, that it was nonpartisan. They were just reproducing nearly century-old propaganda that formed a core part of what our academic grandparents had been taught when they were students.

In reality, in both the 2010s and the 1950s, racial totalitarianism lurked as a subtext to both science's uses and claims about American democracy. The impact on Vietnam of the chemical weapon napalm, which was developed at Harvard by chemist Louis Fieser, is widely known. The long-term health effects of the herbicide Agent Orange, which effectively worked as an anti-civilian chemical weapon in Vietnam, continues to impact families across Southeast Asia and the United States, including my own. Also, there were the incarcerated people in Holmesburg Prison, many of whom were Black, who were early test subjects for the scientists experimenting with Agent Orange. Arguments will continue endlessly about whether the US ever would have dropped nuclear weapons on a European nation like Germany, while it readily dropped them in Japan, despite the fact that Japan was likely already going to surrender. From Dr. J. Marion Sims's nineteenth-century gynecological surgery experiments on unanesthetized enslaved Black women to the Tuskegee Experiment on Black men and their families to careless uranium mining on Indigenous reservations and nuclear weapons testing that left Pacific Islanders to live and die with the nuclear fallout, the devaluing of Black American lives and Indigenous lives around the world has played a key role in scientific experiment and development.

These experiments were carried out because of a racist refusal to value people's lives, but the US also has a history of using technology to directly attack people of color freedom movements. Deploying the technologies of increasingly capable airplanes and bombs, the US military used technological prowess to intervene in global anti-colonial movements all over the Global South—from Africa to Asia to Latin America. Moreover, surveillance technologies, such as listening devices, that often rely on solid-state physics discoveries, have long been deployed against Black civil rights organizers, from the Black Panthers to members of Black Lives Matter. Some of us in the scientific community have witnessed the damaging impact of scientific technologies on our own families, for example, because so many of us know and love veterans of the wars in Vietnam and Iraq. We may not have blamed the institution of science for what happened, but it is unchecked, unethical uses of science that have played this damaging role. It is undeniable that global warming, which has already taken many lives, is an example of technological advancement: we have developed the capacity to warm an entire planet. This might be useful for those thinking about escaping to Mars, but for those of us with any kind of attachment to Earth, it is a technological disaster.

Unfortunately, for a long time, the effects of global warming primarily impacted people in the Global South, whose scientists typically struggle to gain institutional and political power and then use it to counter the abuses of the Global North. Global warming lurks in the background, but until the last few years when hurricanes and blizzards really started to wreak havoc on places where large numbers of white people live, there was little political will, even among liberal politicians and their constituencies, to recognize the fierce urgency of responding. As I wrap this book, the impact on those of us in the Global North has become more evident. My home state of California has been on fire for weeks, and every day, I check the news to see if the Mt. Wilson Observatory—where Hubble first observed the expansion of the universe—is still standing as the Bobcat Fire rages in Angeles National Forest. My family and friends in California, Oregon, and Washington have to check the air quality daily to see whether it is safe to go outside, and my mom and stepfather were asked at one point to prepare to evacuate. Thousands have lost their homes, their lives changed forever. Being homeless is even more unsafe than it was before. Although we are starting to see relief in the west, my adopted home state of New Hampshire is implementing restrictions on water use because of a severe drought and our own forest fires. Global warming is here, and there is no escaping it.

In this context, scientists who are stunned by Donald Trump's rise to power and newly afraid of the rise of totalitarianism should ask some pointed questions about how they had previously ignored the role science played in protecting the institutions that facilitated racial totalitarianism. Were these same scientists outraged that the precious metals that they use in their experiments, or are used in the experiments that undergird their work as theorists, are often collected unethically? From the lands of the Navajo Nation to lands all over the continent of Africa, waterways have been poisoned and people have been kidnapped into slavery, all in service of collecting the materials that undergird modern technologies. Where has the collective outrage been? Of course organizations like Science for the People, the Union of Concerned Scientists, and the Bulletin of the Atomic Scientists have made some attempts to change the status quo, but they have never been mainstream professional societies and have never been seen as central to institutionalized scientific discourse.

It is also easy for those outside the United States to point the finger at the US as a singular, evil global juggernaut, but the historic Greenham Common Women's Peace Camp about sixty miles east of London, which I visited multiple times as a child, was formed in direct opposition to British complicity with nuclear weapons development at the Royal Air Force Greenham Common. French uranium mining in the African nation of Niger began in the 1950s, and today the top producers of uranium are Kazakhstan, Canada, and Australia. The US is hardly alone in its uses of colonized lands to look for materials that can serve technologies of death.

This, in the end, is science under totalitarianism. As Commander Sisko says in the series premiere of *Star Trek: Deep Space Nine*, "I live here." We live here, in a society that has long been governed by racist and colonial totalitarianism. For a long time, institutionalized science, which has traditionally been exclusively by and for white men, at best ignored this reality, and at worst played willing foot soldier for the cause. To be anything different, we must start where we are and talk about whether this is who we want to be and the world we want to make because no matter what Dan Savage coos at you, it doesn't always get better. Progress is not a guarantee. This should be the definitive lesson of November 2016 for anyone who is new to this understanding of the world. We have to bend the arc of the moral universe ourselves because the moral universe is populated by people. People are the ones who need to shift.

Importantly, they have to shift in the right direction. Clearly some—Trump voters, the Democrats who champion centrism and neoliberalism—are shifting in the wrong direction. Now that the inevitability of progress has been proven a lie, it is time to confront the colonialism and anti-Blackness and xenophobia that are foundational to what America is and why it exists. Now is the time to confront the history of gender and misogyny and how enforcing a gender binary in an essentialist manner was part of the violent colonial project. We must accept the truth about ableism and its connection to our fierce belief that some people are disposable and worth less. All of these things are tied together under the tent of white supremacy. We must somehow hold all of these actual truths about America to be self-evident. We must be humble enough to hear the lessons of people who are critiquing the colonialism of institutionalized science.

FOURTEEN

BLACK FEMINIST PHYSICS AT THE END OF THE WORLD

> The islands emerge from the depths, from the darkness that precedes their birth. [Leilani] Basham argues that, similarly, political autonomy is a beginning of life.
>
> —Noelani Goodyear-Kaʻōpua, "Introduction," *A Nation Rising: Hawaiian Movements for Life, Land, and Sovereignty*

THERE IS BEAUTY AND POWER IN BLACKNESS. MY ANCESTORS survived the Middle Passage. That's not some kind of guarantee that I will make good choices. It's not biological determinism either. Instead it is a hint, a guide. I was raised on Black Power, and I know I have to return to that idea whenever I begin to feel hopeless. My grandmother Elsa, losing her voice on her death bed in Brooklyn, whispered over and over again, "My granddaughter. I left Barbados on my birthday, to come here." Every day the ancestors whisper to me, "Survive, survive, thrive, and take us—and all you can—with you."

I don't consistently experience a faith in my ability or in our collective capacity to do this, because though many of us are no longer in bondage, some of us still are, and the collective species of Earth are not OK. We as a humanoid species are not OK. So many different combinations of "we" are not OK. Some of us bear the burden more than others. The whales did not cause the oceans to warm and the fish to die off, but they must live with the consequences. Most of the humans on the planet are not responsible for this terrible phenomenon either. A small number of us have contributed a lot more than others. Even fewer of us had extensive power to decide about the nature of our contributions. Some of us had more of a choice than others. Those who had less choice and made the fewest contributions will also suffer the worst, most devastating consequences. Americans witnessed this in 2005 when Hurricane Katrina ravaged the poorest and often darkest people in Louisiana and Mississippi. But we will all feel it. We will all lose things we loved and enjoyed. Earth won't be Earth to any of us again, if those in power are not careful—unless those who are disempowered take control out of the hands of the dangerous few.

We are not OK, and now is a strange time to publish a book about what an amazing and fascinating place the universe is and the significance of physics as an analytic framework. We are on the precipice of something spectacularly awful. We are already in the regime of something irreversible. This book has, in part, been about what science looks like under totalitarian white supremacy. But my dear friend Ayinde Jean-Baptiste, a lifelong activist, once told me, "Let's remember what we're for, and not just what we're against." That's a lesson that I carry into every room. I tend to be against a lot of things because so much of what goes on is simply not OK. But I want to have a vision for the future too, one that is flexible and that understands, as Alexis Shotwell tells us in *Against Purity: Living Ethically in Compromised Times*, that impurity and imperfection are impossible to avoid. I want to be forward-looking and adaptable so that I can work for a world where humanity—as I know and love/hate it—can survive and thrive, rather than end up totally consumed by ecological catastrophe.

I'm a scientist, meaning my whole job is predicated on the idea that there are still questions that I don't know the answers to. Just as Adrienne Rich once said about art, there is no simple formula for the relationship of science to justice, and I cannot provide a simple prescription or algorithm for linking them. But, like Rich, I do know that science—in my own case the work of studying the origins and history of the universe—"means nothing if it simply decorates the dinner table of power which holds it hostage." I have a responsibility to refuse to assimilate into a scientific culture that assumes white supremacy, capitalism, colonialism, and militarism are simply the cost of doing business. I also know that we are approaching the end of the world, and if we are to salvage life as we understand it and work to avoid repeating the sequence of events that led us here, we will need a new way of thinking and being in relationship with each other.

Part of my science has to be rethinking not just how academia works but what I want science to do in the world. It is here I turn to Black feminism, partly because it is my cultural and intellectual heritage. I was raised by a Black feminist, and so much of how I see the world is through the framing my mother taught me. Black feminism is an analytic that centers Black women's intellectual thought, Black women's experiences, and the specific way that race and gender as social frameworks operate together in the lives of Black women. Black feminism as a mode of thought has been organized around responding to racism and sexism and the unique way they combine in Black women's lives, misogynoir. I choose to focus on Black feminism in particular not out of nationalist sentiment or because I don't recognize the influence that the feminisms of other women (of color) have had on me, but rather because as a Black woman it is my starting point for developing an analysis. Black feminism has also shown the necessary flexibility to expand and include femmes, nonbinary, and agender folks, and it illuminates all kinds of social structures that impact *everyone's* lives.

Importantly, misogynoir has often denied Black women and femmes a seat at the institutional knowledge production table, but as Patricia Hill Collins outlined beautifully in her book *Black*

Feminist Thought, that hasn't stopped us from thinking or creating ideas. You just have to look past traditional methods of record keeping to find this history and legacy. We have to look for a different context too. In a discussion of the politics of Black women's idea-making, Collins cites Black feminist novelist, poet, and thinker Alice Walker's *In Search of Our Mothers' Gardens*. Walker writes, "I believe . . . that it was from this period—from my solitary, lonely position, the position of an outcast—that I began to really see people and things, really notice relationships . . . the gift of loneliness is sometimes a radical vision of society or one's people that has not been previously taken into account." Re-reading Walker's words after the publication of my scholarly article on white empiricism, I realized that I was the person who wrote that white empiricism article specifically because of the kind of loneliness produced by being a politically radical Black woman in physics. With the perspective of someone who wasn't necessarily keen on being an insider, I was able to see structures that were invisible to people who were.

Physics has a culture, and it is deeply entangled with power and inequalities in its distribution. Although the physics community is purportedly global, Western hegemony is strong. Exemplary of this culture is a myth I hear from scientists and science enthusiasts: that science and progress are synonyms for one another. And yet, as I said in the last chapter, global warming is a technological advancement. And yes, it is indicative of our capacity to change the world around us. But in the context of Earth's fragile ecosystem, it's awful. It's not a positive development, and our world is not OK. Species are dying off, and parts of the planet are becoming increasingly uninhabitable.

And beyond the immense physical and psychological suffering caused by these changes, there is also the immense physical and psychological suffering that began with colonialism and is continuously produced by white supremacy. The technological advance of global warming is hardly progress when we consider the way it is enhancing colonialism-induced inequalities between human beings while wiping out the diversity of Earth's ecosystem. Western science imagines the world in hierarchical terms. It is a way of controlling the universe; we make predictions, we break things down to their most basic elements, we know things intimately. Even for all of its facts, Western science has struggled with acknowledging a core reality: humans are not the masters of our ecosystem but rather dependent on it. We need it more than it needs us, but imperialists and settler colonial states have behaved as if the reverse were true. This core mythologizing about humanity's relationship with the planet has produced an ecological disaster that is holding Earth's living creatures hostage.

I want our existence to be built out of love for a better, different world. The sun has about five billion years left before it destroys Earth, but it's hard to imagine that our species, which has only been around for a couple million years, will last that long. How will we ever be OK when we are facing a mess that is this big, while the people who made the mess still have most of the visible power? We have to shift who holds the power—and we have to move to a framework that centers good, life-honoring relations between people. While the Enlightenment may have helped lay the foundation for the way that I see the world in my day-to-day science, it did not leave us with a good legacy on valuing human life. We must start looking elsewhere for a new way of looking at the world of relations between living things. It may be that in tandem with this, we will find that there are new ways of seeing the universe itself. We may find that it gives us new reasons to care about where the universe came from and how it got to be here.

I did not always have this mindset about radically shifting how we saw the world. A few years ago, I was invited back to my PhD institution, the University of Waterloo in Ontario, Canada, to give a talk on diversity. That evening, I had dinner with two student groups that hadn't existed when I was there: FemPhys and the Black Association for Student Expression. I was extremely thrilled to see how the conversation about Blackness and feminism at Waterloo had grown by leaps and bounds in just the five or so years since I had left, and grateful to the Waterloo Public Interest Research Group for paying me to come. When I asked the students in FemPhys, who represented multiple minoritized gender identities, why their organization was called FemPhys rather than the usual women in physics, they told me it was because they weren't a women in physics group but rather a feminist physics group. That distinction marks what I see as an essential shift toward a political commitment to feminist analysis and action.

I share this story because I consider myself to be a Black feminist physicist now, and part of the journey to that understanding includes a group of primarily white physics students who were committed to imagining a different and better way of doing physics. They reminded me of the email signature of the best professor I had at Harvard, Adams House Dining Hall chef and UNITE HERE Local 26 steward, Ed Childs. Childs always signed his emails, "A Better World Is In Birth." I had always oriented myself toward believing that this would happen through labor and radical grassroots organizing. I hoped that would one day force physicists to do better by the world. It had never occurred to me that reimagining physics itself could be another starting point. Even though I was well known by then for my critiques of how the physics and astronomy communities treat minoritized people, especially Black, Latinx, and Native students, I was still thinking in two minds.

In one mind I thought about diversity and shifting people toward anti-racism. In another, I thought about how capitalism harms people, how physicists often help capitalism do this harm, and how Black physicists are sometimes those physicists helping capitalism do the harm. I consciously thought about keeping the part about Black physicists separate because I know, intimately, how hard it is to be Black in physics. I know nearly every Black physicist feels they have worked a miracle by surviving in this world. It's hard for me to knock them for the research choices they've made, because I know our families are looking at what we do and asking us to be successful, by any means necessary.

In the years since, it has been harder to hold the idea that Black physicists shouldn't be accountable for our professional choices because the real problem is supposedly violent white supremacy, which is largely enforced by white people through their participation in whiteness. That's still true, but it's also true that we do have a choice about how we participate. We do have a choice about whether we are part of anti-Indigenous activities in astronomy and about the extent to which we try to avoid complicity in America's imperial wars. I want Black physicists to help lead the way. I don't want us to integrate into an astro/physics community that has historically been part of the problem. My thinking about this is inspired by people like Sudanese materials scientist Maram Ali Ahmed. While listening to a YouTube interview with her about her work on water filtering technology, I realized that to the extent that science can improve and add to our lives, we must be allowed to advance its progress on terms that support the thriving of our home communities.

In some sense, this practice is as easy and big as asking: what are the conditions that our communities need to see the Milky Way? This might not sound like a Black feminist thought, but it is one that asks about looking at the night sky from a particular organizing perspective that I come to because of my upbringing and training in Black feminism. It may not sound like a scientific question either, but as I described in earlier chapters, a lot of work in observational astronomy is completely driven by the question of what we need to see the sky. My first experience with good seeing was in 1996, when Comet Hyakutake visited our skies. My mother made a point of driving us all the way to Joshua Tree National Park so that I could see it. By then, she had made sure I had a small telescope, which I rarely used because what I could see with it in Los Angeles skies wasn't terribly exciting to me. We took the crappy binoculars I had because those are actually better for viewing a quick moving object like Hyakutake. The telescope and binoculars cost money, probably both gifts from my dad at my mother's request. And driving to Joshua Tree cost money and took time, two things my mom didn't have much of. Looking back, I understand that my mom must have made difficult financial choices to make that trip happen, especially since at the time she was recovering from a terrible bout with what would come to be a chronic illness.

That night I must have seen something of the Milky Way and not known what I was looking at. My mom and I barely knew how to look for the comet. A year later, I saw some of the Milky Way from Sequoia National Forest during a camping trip. I remember saying to my dad, "Why does the sky look like that?" And he responded incredulously, like I should have known, "That's the Milky Way." Although he had grown up poor, he had grown up with a far less polluted sky than I had. And what I saw in Sequoia was really only a hint of the possibilities. It wasn't until the trip to Chile in 2011, almost fifteen years later, that I would get a sense of what the whole thing looked like.

My Black feminist thought about this story is: what are the conditions we need so that a thirteen-year-old Black kid and their single mom can go look at a dark night sky, away from artificial lights, and know what they are seeing? What health care structures, what food and housing security are needed? What science communication structures? What community structures? What relationship with the land do they need? And I do not mean to ask these questions on behalf of a child who has been marked as highly gifted and who is confidently planning to study astro/physics at Harvard or Caltech one day. I mean any thirteen-year-old Black kid. A Black feminist physics requires asking these questions and understanding that there are a whole series of human-made structures that interfere with the night sky, not just passively, but actively, aggressively.

When I asked leading New York–based educator Erica Buddington what she thought our communities needed, her answer was very specific: "Easier routes to Long Island and upstate NY. Robert Moses created our highways. He made low bridges so city buses couldn't use the freeways." I was particularly struck by this response because it gave me context for my mother's post-immigration upbringing in Brooklyn. And it reminded me that despite the ways we struggled financially, my mother had a car. Los Angeles, where I grew up, remains infamous for its terrible public transit infrastructure. In many ways, you are trapped if you don't have a car. Public transportation is part of the answer to my Black feminist astronomy question too. This is different from the response most scientists would leap to first: "outreach programs." And we have those. We have scientists who helicopter into inner-city schools with the promise of opening doors for students. But those programs don't feed the kids three meals a day and make sure they have stable housing. Importantly, they rarely provide support for the whole family, and at the end of the day my question isn't about the kids getting opportunities their families didn't. It's about lifting all of us up, together.

"Us" also has to include the Joshua trees too. In March 2020, I returned to Joshua Tree for a night of stargazing and some hiking among the trees. Afterward, I re-read the press about a study from 2019 that showed that even if we take dramatic action to slow down global warming, by 2070, the Joshua tree population may be 19 percent of what it is today. Without dramatic action, these magnificent trees that have been here for 2.5 million years will be gone. What life will hold the Mojave Desert's land in place, making it a nice site for stargazing without getting hit by nasty wind storms? The right to know the night sky involves a right to life for the creatures that live closest to the land, like the cholla cacti, Joshua trees, giant and coast redwoods, and other flora and fauna that populate national parks and Tribal lands.

I guarantee that the fight to make the night sky accessible will require personal choices that will feel like enormous sacrifices, sacrifices that may not feel like they pay off in our lifetime. Scientists of my generation—those of us who have survived the brutal filters at the myriad training levels—are used to begging for money but are not at all familiar with paying the costs associated with social justice struggles, including lost positions and maybe lost careers. Here and there one can find stories of women who sacrificed professionally to hold a rapist accountable, but generally speaking, we've all learned to go along in order to get along. Even I fall into this category to some extent. I take more risks than most and have rarely chosen to stay silent when I had the opportunity to speak truth to power, but it's also true that I have made those choices under the constraint of trying to maintain some semblance of a career. I have chosen relative silence about how the military shapes our diversity discourse, in particular the way it has funded the minority professional societies that I have devoted so much time to.

In the end, Audre Lorde's reminder that our silence will not protect us is for scientists too. And this is how the totalitarian state works on scientists. It offers its valued brains a sense of stability in exchange for complicity—in exchange for silence. During the time I wrote this book, billionaire Elon Musk's company SpaceX began to blight the sky with Starlink satellites (and Jeff Bezos's Amazon got permission to join them). Astronomers who were used to being the beneficiaries of innovations in space technology, and who have generally celebrated SpaceX's growing power in the American human space program, are furious. There are so many satellites in the sky that the telescopes that cost billions to build, collectively, now produce images with satellite streaks in them. Musk argues that Starlink will deliver "development" to far flung places by making internet available to them. But at what cost? And why should we believe him? Increasingly, the night sky that people in those far flung places have come to know and love will be filled with Musk's vision of development—because he's exorbitantly rich and that matters more than democracy to our nation-states. Astronomers have been left begging this billionaire, and his company that is heavily subsidized by American tax dollars, to effectively only do things they like (launch rockets with supplies and astronauts to the International Space Station), but to leave their majestic sky alone. The irony of this in the face of how astronomers have treated Native Hawaiians is a bit much, really.

My work as a scientist is not meaningful if it accepts the world that is or that scientists have a particular entitlement to get what we want, while nonscientists do not. And that is one reason this book opens with stories of wonder that I hope will inspire readers. The universe is a fascinating, wonderfully queer place, and for millennia, trying to understand how it works has inspired humans to stretch their imaginations in ways that we didn't know were possible. I once asked my mother in a moment of doubt what the point of my work was, and she said that people need to know that there is a universe beyond the terrible things that happen to us. The stars, the Standard Model, the way spacetime bends—this way of seeing the world is one that can be inspiring.

But what is inspiration if there is no joy to accompany it? I remember the first time I saw two physicists yell at each other during a "talk"—the colloquial term scientists use for an oral presentation on one's research. This one particular talk was being given by a researcher who supports the inflationary model of spacetime in the early universe. Another researcher who has constructed an alternative to the inflationary model happened to be in the audience. As a graduate student, I found talks confusing—I still don't learn as well from listening as I do from reading—so I don't remember the ostensible reason why the yelling started. And in fact, it wasn't the first time I had seen male physicists get angry with each other. My first year in my PhD program, I was sitting with one of my advisers and another distinguished researcher in the field, watching a younger professor informally present a radical and unpopular idea at a chalkboard. I don't know what the distinguished professor said, but at some point the younger man started punching the chalkboard and then walked away, went to his office, slammed the door, and screamed. That was more one-sided than what I witnessed at the talk though. In the talk, they were yelling at each other, angrily, as if lives depended on it.

Lives did not depend on whether either of them was right, and I've many times since seen people get really nasty with each other over scientific disagreements. Of course, anyone who has lived in the Boston metro area knows that people get into all sorts of fights over the most ridiculous things—just watch a Bruins game sometime. But the anger is surprising given how sober and somber theoretical physicists usually are when they present their work. Perhaps this stands out to me because of the way I was recruited: with all the excited fanfare that goes into science popularizations. "Isn't science cool? It's so cool! Let us show you how cool it is!" The actual practice of science feels so distant from what we are expected to tell the general public. In private, we are worried, anxious, yet maybe also deeply in love. We don't smile as we present our work, but rather wait for the famous question in the form of a comment—someone raises their hand, says something that indicates they don't understand, but rather than asking for clarification, tells us we are wrong.

I don't understand why we can't be more joyfully loving about the work. I also completely understand it as a Black woman professor. There is nothing joyful about most committee work, about dealing with people asking me if it's OK for the white people to say the N-word (no it is not), about knowing that people will regularly question my competence just because of my appearance and the sound of my voice, about the way universities are increasingly run like for-profit corporations, rather than important public goods. But shouldn't we, especially in the face of that, try to preserve a sense of joy and pleasure when we are talking about wondrous things like quark confinement and the outrageous ways that spacetime gets bent by dark matter?

People often associate the word "pleasure" with sex, but what I'm talking about is having fun while feeling the nonhuman universe out with our brains. People often talk about the importance of enjoying the process, but how can we transmit this to our students and colleagues if we never seem to be enjoying ourselves? I feel sometimes that being geeked is hard because my colleagues are so ready to tell me why an idea is wrong, how it can never work, how their idea is better. As theoretical physicists, we have a distinct opportunity that we should take advantage of more: we are allowed to have lots of ideas, many of them are not immediately testable, and we can just keep making things up as long as the idea hasn't been ruled out by observation or experiment. Why don't we relish that more?

I am especially interested in responding to this as a queer Black femme because the question of pleasure for people like me is so complicated. The academic community has traditionally not believed in the existence of Black humanity, much less the value of Black joy. Historian and women's studies researcher Rebecca Herzig has in fact written a whole book, *Suffering for Science: Reason and Sacrifice in Modern America*, about the significance of suffering as a part of professionalized science in the United States. She traces this way of being with science back to America's puritanical roots. So the problem isn't just not believing in Black joy, but also what happens when a whole professional community doesn't believe in everyone's humanity, or in joy for anyone. The politics of suffering that pervades scientific work produces a lack of social imagination. Our planet is suffering under the white supremacist, colonialist, and patriarchal politics of suffering in science. People are suffering enormously. Inequality, rather than shrinking with the advent of automation, is expanding like a terrible, unbounded economic spacetime—this time pushed outward by greed rather than gravity.

We face difficulty in confronting this suffering because academia teaches its denizens that solidarity is an optional privilege, but we cannot afford a politics or a science that lacks solidarity. In my experience as an academic union organizer, scientists are the worst about getting involved and the most likely to refuse to join the union and to cross a picket line. As the most economically powerful intellectuals in academic institutions, scientists are making a power statement when they choose to ignore a politics of solidarity in favor of getting theirs. This can be an especially tough pill to swallow for minoritized people like Black women because we often feel that we have clawed our way into the room. We have often fought racist classmates, sexist professors, and homophobic lab partners alongside the self-doubt that is engendered by a community that is extremely hierarchical and obsessed with "meritocracy" but primarily constructed out of straight cis white men always having more of an opportunity than everyone else. It is hard, in that context, to accept a lens through which we are people with power. Yet, as scientists, here we are. As graduate students, we are more likely to be paid better than the nonscientists, including our colleagues doing radical feminist and Black studies work. As faculty, our starting salaries are higher and the grant money to support our work is far more extensive—by orders of magnitude. This translates into disparate treatment by our academic institutions. Scientists wield undue and disproportionate power.

Black and Indigenous feminist thought offers lessons on how to respond. These modes of seeing the world can help us understand what it means to be in solidarity—be in better power relations—with each other. As the "Combahee River Collective Statement" articulated it, "This focusing upon our own oppression is embodied in the concept of identity politics. We believe that the most profound and potentially most radical politics come directly out of our own identity, as opposed to working to end somebody else's oppression." Tension between this articulation of identity politics and a politics of solidarity is apparent, not real. I do not believe that the Combahee advocates for the promotion of Black women's concerns over the concerns of others but rather for understanding what can motivate our organizing—including our solidarity activities. We can be autonomous in both our decision-making and thought as Black women (and femmes and trans men and other gender minorities) while understanding that the great lesson of our ecosystem's deep troubles is that we must take an "all of us or none of us" approach. Black American women will not be OK if non-Black Indigenous women are not OK. We may feel like we're getting somewhere, but as long as the system that creates racial and ethnic hierarchies is in place, it threatens us all.

Traditionally Black feminism has emerged in connection with the urgent sociopolitical needs of Black women and their families—for example, confronting southern Jim Crow and northern segregation. Science beyond the social and life sciences—i.e., psychology, medicine and public health—has not traditionally been the standpoint from which Black feminism has emerged. It is also not a standpoint that Black feminism has always openly and consciously attended to. This is not to say that physical science hasn't been part of Black women's thought. Harriet Tubman used the North Star to run away from slavery, and my guess is that it wasn't the last time she used celestial navigation to lead people to freedom. I imagine that enslaved Black girls, just like white boys and girls, looked at the night sky and thought about the patterns they saw. As a descendant of enslaved people, I am often grappling with a kind of grief about histories that are hidden, some of which will never be recovered. Although the legacy of enslaved Black women as midwives, health care workers, and agriculturalists is increasingly recognized in academic literature and popular knowledge, the idea of a Black woman engineer or physical scientist still seems like a primarily twentieth- and twenty-first-century phenomenon, one that begins around the time that the so-called hidden figures began their work at NASA.

Part of my intellectual task as a Black feminist physicist is to try to understand what it might have meant for Black women to engage what we could classify as "physicist thought" during a time when Black women were so dehumanized by the law and formal educational systems. I don't know how to begin looking for those women in the archives, but part of my Black feminist physics is disrupting the idea that Black women need to learn a new cultural mode to be creative thinkers who can have an impact on physics. Joseph Martin writes in his book *Solid State Insurrection* that physics is what physicists say it is. Our understanding of our discipline is socially defined. I don't know what vocabulary these physical science thinkers thought in. But I know they must have existed, the way I know that Harriet Tubman must have had a special relationship to the night sky.

Of course, it is not enough to repopulate history. Not only should we ask where the hidden figures are, but also what they were doing and what agendas they were serving. We must reimagine physics through a Black feminist frame. This too is my task. It does not have to be my task alone, and it should not be a Black women–only task. Operating from a framework that will improve the lives of Black women means operating from a framework that will improve the lives of many people. Social justice organizer Dean Spade once said in a speech that if we center the most vulnerable in our movements, everyone's lives would get better. Black women, especially Black trans women, and Indigenous women, as well as Two-Spirit and nonbinary folks, are some of the most marginalized people in this white supremacist heterocispatriarchal society. If we can organize ourselves using principles from Black feminism, Indigenous feminism, and other women of color feminisms, we can create communities that are in good relations with each other. We can also be individuals who are in better relations with ourselves and each other.

Black feminist thought does not mean just engaging with Black women or other Black women's ideas. It is constructed in part by Black women in the spaces where we are. The space I find myself in, partly, is this academic and specifically scientific one. The discipline of Black feminism and its analytic frameworks guide me as I consider how to orient myself in this space. Science is a space that was specifically built without me in mind; in fact, it was built with my subjugation in mind, more a scene of subjection (per Saidiya

Hartman), than a scene of liberation. For many years I did not ask myself the question of how physics could play a role in liberation. Superficially, as a teen, I imagined that a holistic theory of quantum gravity would allow us to calculate right choices, right thought, right intentions. This is partly because I was reading a lot of Thích Nhất Hạnh my senior year of high school. I am still grateful for and deeply influenced by Nhất Hạnh's lessons, but my interpretation of them had to mature and deepen. The human world, and its relationship to the rest of the universe, is not as simple to understand as the Standard Model of particle physics. Quantum gravity will likely never provide the security and predictive power that I hoped it would. But I refuse to accept that the answer is the other extreme, that physics is only useful insofar as it serves immediate and long-term human material needs—things like energy and water, delivered in and used in ways that are healthy for us and our ecosystem.

But humans can live gracefully through our imaginations. Every community has a cosmology in part because it plays a role in our social ordering; every community also has a cosmology because we want to connect the places we are from with the world we witness into the greater cosmic timeline. As a Black feminist physicist, I want to help people maintain that connection to the world beyond day-to-day material stresses. I want to work to ensure that this connection is accessible to Black children and Indigenous children and to teach them to use the knowledge ethically so that we can be in good relations with each other.

I'm particularly interested in work that we have seen emerging from different organizing sectors, especially within The Movement for Black Lives, the movements that coalesced around #NoDAPL and #IdleNoMore, prison abolitionist work, and yes, antifa activists. From all of these communities, ideas are emerging that are oriented toward what the Global Women's Strike calls "investing in caring, not killing." Abolitionists like Mariame Kaba are teaching us that a politics of disposability does not produce healthy communities. The Movement for Black Lives set a powerful organizing example when it created "A Vision for Black Lives: Policy Demands for Black Power, Freedom, and Justice," which is a whole platform that serves as a guide for creating policy and community organizing. In other words, it targets all levers of power with potential for change in society.

These movements are popularizing and (re-)articulating new ways of thinking about justice. Rather than orienting ourselves toward retribution, punishment, or threats of harm, we can consider instead the idea of transformative justice. Generation Five, a collective that seeks to end child sexual abuse, defines transformative justice as "a liberatory approach [that] seeks safety and accountability without relying on alienation, punishment, or state or systemic violence, including incarceration and policing." When Lacy M. Johnson talks about her rapist living in service of other people's joy, I believe she is talking about what transformative justice might look like for her and the man who violated and tried to kill her.

I have become increasingly interested in what we can learn from anarchist thought. For example, the idea of "mutual aid" moves away from the hierarchical orientation that statism encourages us to have. Big Door Brigade (BDB), an organization that began this work in June 2016, defines mutual aid as "a term to describe people giving each other needed material support, trying to resist the control dynamics, hierarchies and system-affirming, oppressive arrangements of charity and social services. Mutual aid projects are a form of political participation in which people take responsibility for caring for one another and changing political conditions, not just through symbolic acts or putting pressure on their representatives in government, but by actually building new social relations that are more survivable." BDB is not just proposing an idea for someone else to execute but has in fact created what it calls a "mutual aid toolbox" that is available for free on its website. It is helping people implement these ideas.

The movement to protect Oceti Sakowin (Lakota/Sioux) lands from the Dakota Access Pipeline showed us the potential of so many of these practices when fighting for the autonomy, freedom, and well-being of our communities. Standing Rock, along with the Occupy Movement, exemplified the strength inherent in solidarity between communities, as well as the vigor that the state and capitalism's other protectors will show in trying to keep us apart. With Occupy we see also the difficult challenges that we face in solidarity work. Coalition organizing under white supremacist conditions is never easy. Part of the work is about challenging the way we have internalized white supremacist and capitalist narratives about who is valuable, whose ideas should be listened to, and even how to agree and disagree in ways that aren't oriented toward consolidating power in the hands of the few.

Writing under the pseudonym Alexandre Publia, a South African activist implicitly raises this point in the context of anti-racist and anti-colonialist organizing at South African universities when they note in the first issue of *Abolition: A Journal of Insurgent Politics*, "The military-industrial complex necessary for state security and surveillance apparatuses remains supported by universities' physics, robotics, and engineering departments." In other words, physics continues to be dangerously proximal to the people who work to suppress freedom movements, rather than the people enlivening freedom, mutual aid, and good relations. As Fenton Johnson writes in "The Future of Queer," "Thanks to science, we are the first empire in history to possess the knowledge of what we are doing to ourselves, the causes of our environmental self-destruction—though, as prophets and artists and writers demonstrate, the visionary imagination has no need of data to read the writing on the wall." Can we respond by building a community of scientists hell-bent on using our visionary imaginations?

This is how we challenge the capacity of totalitarianism to use the scientific community to do its bidding. Science must be undergirded by a commitment to being in good relations with the world that is to come, and that requires imagination and a sense of wonder at the universe that is. We must move past talk of ethical conduct and toward anti-colonial action. Science needs an anti-colonial code that includes a better understanding of the dynamical relationship between human ideas and the world onto which we project them. We struggle to have this conversation because in a science operating under capitalism, we also operate on what I would call a colonial clock, one that ticks under the command of colonial modes of thought. I actually first came to this realization in thinking about the struggles around the Thirty Meter Telescope and was heartened when I found writing by Dr. David Uahikeaikaleiʻohu Maile that organized and deeply interrogated this exact question in a pair of essays with the title "Science, Time, and Mauna a Wākea." One of the points that Maile argues in his essays is that the Western focus on certain types of technology as "progress" creates a kind of time pressure to do science without considering the impact on people. In the end, we can choose not to acquiesce to this pressure. We can resist. We can choose a different timeline. And we should.

And we know how. I have had to realize how much my scientific training has distanced me from values instilled in me not only by my mother, who was steeped in Afro-Caribbean spirituality, but also the Lakota friends who at points welcomed both my mother and me into community and sometimes also ceremony. I was struck by this when reading Shawn Wilson's *Research Is Ceremony: Indigenous Research Methods*. He quotes a colleague, Lewis, on the use of technology: "This machine here is made from mother earth. It has a spirit of its own. This spirit probably hasn't been recognized, and given the right respect that it should. When we work in a world of automated things, we forget that . . . everything is sacred, and that includes what we make." This should have been a foundational thought for me with everything I touched as a student and early career scientist, but instead, my training as a physicist taught me to forget that, as Wilson goes on to say, "spirituality is one's internal sense of connection to the universe." Because I did not feel a faith in the supernatural, and I was not in an environment that respected a sense of spirituality beyond Western conceptions of religion, I nearly lost my connection to a very basic, spirit*ed* relationship that I had to the universe. I am grateful to the Maunakea protectors whose urgent work pushed me to rebuild.

In "The Point of Science: Lessons from the Mauna," I discussed one of the lessons I have learned from the protectors: the kanaka idea of pono science. Importantly, I didn't share this lesson for non-kanaka people to co-opt this idea. Rather, I believe it's important for all of us to see that we have contemporaries who are imagining other ways of being and other ways of knowing. I am against molding Indigenous ideas to suit the needs of consumptive colonial civilization or even other Indigenous communities in need of a rubric. I believe others—those who could be identified as settlers/the planned and primary beneficiaries of settler colonialism—must root themselves in respecting the autonomy of our Indigenous communities. I believe those who, like me, cannot trace their heritages back to the Indigenous peoples that form our roots, must organize ourselves to find similar ways to be in good relations with the land and its family, the people, including ourselves. These North American and Polynesian frameworks like pono remind me of the importance of learning about articulating analogous frameworks that arise in the communities I come from, including abolition as an ongoing process, not one that ended in 1865. This may be difficult to accept for those of us who are descendants of Indigenous people who were forcibly severed from their land and heritage. Members of the Black Atlantic diaspora can feel like lost people, wandering in the desert without a homeland and without a language.

But we have not lost as much as we think. Our African ancestors found ways to pass their stories and parts of their language, food, and spiritual traditions down to us. We in turn are tasked with the hard but wonderful work of growing Afrofutures from the seeds they left behind. I believe part of this work is enacting practices of solidarity with those we understand to be a part of our immediate community, as well as those beyond it. Indigenous scientists, including those of us who are part of the Black diaspora of the Americas, will gain enormous strength as we learn to think with each other and enact forms of solidarity with one another. It is here that we will throw off old colonial precepts like the gender binary and affirm for ourselves that we are modern—that we have never stopped being modern—without giving in to the voices that encourage us to destroy the planet in an effort to prove it.

DEAR MAMA, THIS IS WHAT MY FREEDOM DREAM LOOKS LIKE

To the one and only Margaret Prescod:

I feel most at home scientifically with a Lagrangian. I know I will never be able to explain to you what a Lagrangian is. "Your mother's an idiot when it comes to math," you always say. But the Lagrangian is an equation that is tied to the most fundamental properties of a physical system. There are rules for how to extract information from a Lagrangian. What is the system's future? What is its past? I can gather these pieces of information by calculating the equations of motion from the Lagrangian—these tell us how the system evolves. I can also calculate the energy in a system. This is the Hamiltonian. Technically the Hamiltonian framework is an equivalent one to the Lagrangian, but I don't feel as at home with it.

There are rules for how to write down a Lagrangian. Often this first involves a term to describe the "kinetic energy"—the energy associated with movement. Next you include a term to describe the "potential energy"—the energy inherent to whatever object you are describing. The potential, as we call it, usually has at least a term to describe mass, if the system has a mass. Sometimes it's not helpful to use distinct kinetic and potential energy terms. For example, when describing light, we use a mathematical object called the electromagnetic tensor. This tensor is a box that contains information about the electricity and magnetism that together produce light. Yes, including the light that used to wake me up as a child early on weekend mornings unless my bedroom curtains were thick and closed carefully.

This is what freedom looks like. Freedom looks like being able to think about Lagrangians and how to craft new and interesting ones to solve problems like dark matter and dark energy without worrying about cops killing Black people. Freedom looks like joking around with a fellow cosmologist about how she doesn't feel comfortable with Lagrangians at all and then slipping into a conversation about the latest *Star Trek* series and my annual trips to the Star Trek Convention. Freedom looks like knowing that Black girls—especially including dark-skinned ones, trans ones, and disabled ones—*will* grow up to find freedom in a Lagrangian and the symmetries that govern particle physics, because their conditions make it possible. Freedom looks like knowing that they won't be raped or experience domestic violence along the way. Freedom looks like knowing that they can have children and not face professional repercussions. Freedom looks like knowing that they will survive childbirth at the same rates as white women. Freedom looks like kanaka ʻōiwi autonomy and all of us having access to our ancestral intellectual histories. Freedom looks like the dark night sky and everyone having a chance to look at it, wonder about it, and know it.

Elmer S. Imes, the second Black American PhD in physics, was born in 1883 to Elizabeth and Benjamin Imes, who met as students at Oberlin College. While Benjamin's family had been free since the American Revolution, Elizabeth was born enslaved. The people who purported to own her family, to steal their labor and lives from them, did not intend for her to go to college. Yet she did. They certainly did not intend for her offspring to play a significant role in confirming one of the greatest physical theories of all time, quantum mechanics. They wanted to own her and her children. Instead, Elmer S. Imes was born free, and with that freedom he explored the subatomic world of molecules. He also became founding leader of the Fisk University Department of Physics, a Historically Black College and University department that has played a major role in the history of Black physicists, helping to create a community that fostered the generation I am a part of. I know also that Imes probably had limited options—predominantly white universities did not really hire Black faculty back then. Even so, Imes created freedom for me and many others when he created room for us to think physics. There was—and is—freedom in making room for Black curiosity.

You once told me, "People need to know that we live in a universe that is bigger than the bad things that happen to us." I hope this book helps to fulfill this goal, which you sacrificed so much to help me achieve. I think about that Grateful Dead line, "what a long, strange trip it's been." It's your fault that I can quote the Dead at length. I don't think I've ever told you that. There are many conversations we haven't had and may never quite know how to have with spoken words. It wasn't until I moved out that I learned that I have an easier time communicating with the written word. And when I started my blog *The Disordered Cosmos* in 2007, I still believed that what I wanted was inclusion in physics. I thought that was what freedom would look like.

Back then I was still afraid of what might happen if I grew up to be like you. Fiery, not always well appreciated by people who are afraid of fiery Black women, and sensitive, not always around people who understood that Black women can be both fiery and sensitive. I'm old enough now to see that I have grown up to be a lot like you, and that's where some of my best qualities have come from. I also now understand something of how hard it is, although in substantive, material ways, I have it easier than you. I often find myself returning to this comment to the *Los Angeles Times* about you from Walter Gordon III, "Anytime you have a black woman take a position and get noticed there's going to be conflict. She's a powerful person with a charisma, so she gets called aggressive, pushy, confrontational. It's the same old hysterical female trip. So what? She gets things done." As I grow to be more and more like you, to understand the contours and nuances of your organizing decisions and strategies, I understand that inclusion isn't liberation work and that liberation work is often lonely and even within "the movement" sometimes unpopular. I'm also becoming aware of important questions: what am I organizing for? Who am I organizing for? Who do I want to organize with? What do I want us to achieve?

I have often told people, when bragging about you, that in a different world and another life, you would have been a fashion designer. You're always the best-dressed activist at any protest. You're a committed fashionista, and that's what freedom looks like to you. Some people may think this is materialistic, but the older I get, the more and more deeply I understand your perspective as that of an artist and art critic. You will never lead that life because instead you committed to building a world where lots of little village girls like you might have a chance to thrive rather than merely survive in a world that doesn't believe that Black lives matter. This was your sacrifice, and you made it with love and with few regrets. But it also became my sacrifice, one that I did not sign up for. I think you didn't always understand the impact your choices would have on me. But you wanted to be a mother. You wanted to join the group that you had spent years fighting for. You had a right to want a child. And one of the ways the world was unfair to you was not giving you the conditions in which you could just be an artist and a parent, maybe still a teacher too, whatever struck your fancy.

You paid an emotional price too for this. People cannot imagine that a Black woman can be passionate and fiery and still in need of tenderness. I believe this is emotional housework that—like many Black women—you are owed but have rarely been given.

Watching this as a child who didn't have a grasp of the context and didn't yet know what it is like to be an independent Black woman, all I recorded were the ways I wished you were different. I knew from your work that the world needed to be different, but I didn't consider how much the world's problems shaped my perspective on who you were and how you are. You gave an interview where you talked about the hard trade-offs between being a parent and an activist and the impact you know it had on me. That was really hard to read because honestly, I hate that storyline for both of us. It is true that it was hard on me and that I wish you had made different choices. It is also true that I don't want you to live with the guilt of that. What is most true is that you deserved a better world, and so did I and so does Haiti and all of the other people and places you struggle for. No one should have to choose between being the kind of parent they wanted to be; no child should have to give up their childhood. But the world we came into is this one, and it wasn't fair.

There isn't a neat conclusion to be drawn here. That's just the shit of this world. Capitalism, sexism, racism, colonialism? They damage our ability to parent the way we want to and to have the childhoods that would feel most fulfilling. As a child, you watched your dad beg at the big house for money for food. I never had to watch you do that. Instead I answered the phone when creditors called and sometimes ate my best friend's family out of house and home. But I never went hungry. I never had to eat sugar cane instead of a meal. Even though I qualified for free lunch, when you saw that the food didn't fit my unusually picky tastes, you made lunch for me every day. You couldn't afford it, but you fed me. You went into debt buying books about geography and art for me when it became clear I needed more stimulation. You tried to give me a taste of what you knew rich kids had, so that I would have a chance to fulfill my own potential. And while you did that, you sacrificed your own artistic impulses to build a world that looked beyond the politics of thinking only of your own child.

Now I think about what might have happened if you had capitulated to the men (and women) who told you to stop being so aggressive about pursuing the so-called Southside Slayer, who would come to be known as the Grim Sleeper. I firmly believe you saved Black women's lives when you walked the streets letting people know about him, asking people for information about him. I think you saved the lives of the women you spoke to—and I also think that in the long run you are one of the reasons he was caught and forced to stop kidnapping and murdering Black women. You're a real life hero.

You have been incredibly supportive of my scientific inclinations, never pushing me to be more radical, although sometimes not standing in the way when other people in the family did. But the truth is, I understand now that I will not have done enough, by my own metric, if I stop at what is a superficial and capitalist politics of representation in science. I want Black women who enjoy science to live their best lives—not because their interest in science makes them special but because they are, because they exist.

This standpoint has created challenges for me in finding comrades. I've spent a lot of my career as a scientist trying to organize with other scientists who weren't really organizers. Reading that, it sounds like I was confused, and perhaps I was. I was also earnest and hopeful. I was also afraid of what might happen if I said what I really thought, if I laid waste to my carefully crafted "diversity in science" persona. As I got older, that cracked, but I had a hard time taking full responsibility for what I meant by it, trying to keep my worlds separate. I'm ready now to admit that I cannot live as multiple people, and I am the child that you raised me to be.

Thank you for showing me what freedom looks like. Thank you for working an evening job as a secretary so I could graduate from a high school that provided good college admissions opportunities. Thank you for how you responded when I called you during my frosh year and said I was quitting physics. You knew that the world was going to tell me repeatedly that I couldn't be anything because of who I am and who my people are, and it was an important early lesson: it's all hard. Since there's no escaping that, I might as well do what I love.

Thanks, Mom, for making sure that I always held fast to my dreams.

Love,

Chanda

יְהִי רָצוֹן מִלְּפָנֶיךָ ה' אֱ-לֹהֵינוּ וֵא-לֹהֵי אֲבוֹתֵינוּ, שֶׁתּוֹלִיכֵנוּ לְשָׁלוֹם וְתַצְעִידֵנוּ לְשָׁלוֹם.
וְתִסְמְכֵנוּ לְשָׁלוֹם. וְתַדְרִיכֵנוּ לְשָׁלוֹם. וְתַגִּיעֵנוּ לִמְחוֹז חֶפְצֵנוּ לְחַיִּים וּלְשִׂמְחָה וּלְשָׁלוֹם
וְתַצִּילֵנוּ מִכַּף כָּל אוֹיֵב וְאוֹרֵב וְלִסְטִים וְחַיּוֹת רָעוֹת בַּדֶּרֶךְ וּמִכָּל מִינֵי פֻּרְעָנִיּוֹת
הַמִּתְרַגְּשׁוֹת לָבוֹא לָעוֹלָם וְתִשְׁלַח בְּרָכָה בְּכָל מַעֲשֵׂה יָדֵינוּ, וְתִתְּנֵנוּ לְחֵן וּלְחֶסֶד
וּלְרַחֲמִים בְּעֵינֶיךָ וּבְעֵינֵי כָל רוֹאֵינוּ וְתִשְׁמַע קוֹל תַּחֲנוּנֵינוּ. כִּי אֵ-ל שׁוֹמֵעַ תְּפִלָּה
וְתַחֲנוּן אָתָּה: בָּרוּךְ אַתָּה ה', שׁוֹמֵעַ תְּפִלָּה.

May it be Your will, Universe, our Universe and Universe of our stellar ancestors, that You lead us toward peace, guide our footsteps toward peace, and make us reach our desired destination for life, gladness, and peace. May You rescue us from the hand of every foe and ambush, from all manner of punishments that assemble to come to Earth. May You send blessing in our handiwork, and grant us grace, kindness and mercy in Your eyes and in the eyes of all who see us. Blessed are You, Universe, which provides sufficient spacetime for us to realize our prayers.

A NOTE TO THE READER

"We cannot let fear define us in this moment."

—Captain Michael Burnham, *Star Trek: Discovery*, Season 4

THE UNIVERSE IS KIND OF A TERRIFYING PLACE. AS I TEACH astrophysics to my students, I am reminded of exactly how dependent we are on the atmosphere. It holds in the air we need to breathe. We can't do X-ray astronomy from the ground because the atmosphere blocks X-rays from outer space that would be incredibly damaging to our organic bodies. Humans have always known that we are at the mercy of our ecosystems, and as the 2020s unfold we find ourselves increasingly aware of how much we are a part of those ecosystems. We are at the mercy of the worst of our selves. We are all vulnerable, some of us more than others.

We are called to be humble in the face of complexity; we are also called to be brave in the face of catastrophic change. But so much in our lives conditions us to be defined by fear. And these smaller-scale instances of being dominated by fear likely impact our ability to courageously respond to issues that feel bigger than our personal lives. In other words, it is important to practice bravery in our daily lives, in the small things, so that we build the muscle—the stamina—for the bigger tasks required of us to ensure a future for our ecosystems, our family.

Writing *The Disordered Cosmos* required me to reckon with my own sense of vulnerability, both in the face of the climate catastrophe and also in the context of my journey as a scientist. There was so much that I felt I had to get right in the text, lest people say that I deserved all of the terrible messages I had received over the years about how I didn't belong in physics, especially theoretical physics.

This process also made me vulnerable in new ways: some people wanted to call my book a memoir, which felt like an erasure of my role as a science writer. Despite the way parts of me are woven into the text, mostly I wanted to be a vehicle for a message. I did not want to be the story. And, as I say in the introduction, if I seem like the center of the book, you're missing out on what really matters: the question of how we get free and what role a mathematical understanding of our physical world plays in our freedom dreaming.

I knew from the start that a book that takes a holistic perspective on particle physics and cosmology—a book on physics from the point of view of a Black agender woman—wasn't going to sound exactly like books by my childhood idols Stephen Hawking and Carl Sagan. What I hadn't really thought through was that their voices as white men in science were so normalized that people didn't see the biographical elements in their own texts. Carl Sagan's last book, *Billions and Billions: Thoughts on Life and Death at the Brink of the Millennium,* is a collection of intensely personal essays and was in some sense the template for *The Disordered Cosmos.* Yet his book is categorized as "scientific reference" and "cosmology."

Meanwhile, Stephen Hawking's classic *A Brief History of Time* is ultimately a tour through Hawking's own intellectual ideas, but it is not interpreted as a memoir. The text also includes an extensive pop history of physics, but it is never categorized as a history of science, either.

Somehow my book was different.

I refused to write my Blackness, my queerness, and my relationship with womanhood out of the story. Instead, I centered it.

I rejected the idea that good science writing is detached science writing—the idea that, for example, rape is not part of the scientific story. I wrote with the understanding that just because someone is allowed to pretend that they are detached from their subject does not mean they are in fact an outside, disinterested observer.

Indeed, a central premise of *The Disordered Cosmos* is that you cannot take the human out of the science. To write about science at all meant choosing what science to write about according to my own wishes: I was always going to be in the text. A holistic look at physics meant that the need to draw these connections was obvious to me *because* of my social location as a Black agender queer person. Is it a memoir? To the people who came to popular science books for the first time because of *The Disordered Cosmos* or who I am as an author: I concede. All books about science by scientists are biographies of our deeply personal worlds of professional science. And when the books by hetcis white men are called memoirs too, I will be fine with that label.

I was so distracted by the people who would try to box me in that I did not think enough about the earnest reader who came to *The Disordered Cosmos* as a science book that had been written for them. Coming as I do from a British Caribbean labor organizing family, dealing with emotions tenderly is not one of my strengths. My response to problems is to attack them. I am not good at simply holding space for people looking to survey the damage and tend wounds.

Even as I wrote to tell Black(queer)(femme) people that the universe is ours too and to advocate for an end to traumatic STEM education, I did not think of how many of us were so traumatized that the thought of picking up a book that "talks physics" could be terrifying. I was so focused on the importance of writing to folks who had been excluded from the conversation that I did not think about the emotional halo that surrounds that sense of exclusion. I failed to hold space for the reader who was unsure that physics could ever be in good relations with them.

Now I would like to turn to you.

The first thing to know is that there is no test at the end of this book. It does not matter if you know exactly what a quark is. Most physicists don't. You might remember that a quark is a fundamental particle and that combinations of them make up other interesting particles, like protons and neutrons. Or you might not—but now you know where to look in case you ever want to know. Welcome to my life as a physicist. I forget information that I don't need to use on a regular basis. I know how to look information up when I need it.

Science is not about what we know. Science is about what we don't know. A scientist's job is to live at the boundary of what is known and unknown and try to push that boundary forward. That requires being confused—and being comfortable with not knowing the answers because we are confident that we've got a good toolkit. Early drafts of this book emphasized this in places, but I was afraid these notes would disrupt the reader's focus on what I admit is a lot of new and weird information about how the universe supposedly works. The key thing is that if you're feeling confused but intrigued and wanting to learn more, you're having a very "scientist" experience.

Because of the way (STEM) education policy tends to encourage rote memorization, testing, and fixed intelligence mindsets as a pedagogical framework, many people never get to develop confidence in having a good toolkit. I was a relatively good test taker until I wasn't. Luckily I had gotten pretty far (Harvard) by the time whatever relatively untrained organizational skills and social capital I had stopped working for me. Some of us experience being bad at tests earlier than that, and this becomes deeply entangled with our perception of what math and science are. Some of us become convinced not only that our toolkit is lacking but also that we are never going to be able to develop a good one, perhaps because we are dyslexic or have dyscalculia or because timed tests distract us from the material.

If you are someone who has been frightened or even terrorized by science or math, I want to be very clear about one thing: this book was for you. You didn't read it wrong. You may think it went over your head when actually it didn't. And if you were expecting that by the end of the book you would understand things at the level of someone with twenty years of experience in the field, I ask you to stop being so hard on yourself.

You know that dark matter should maybe be called invisible matter and that it's not a particularly good metaphor for Black people. You've heard of gravitational lensing, and you've seen Einstein's equation. You know that the Hubble Space Telescope has produced beautiful images of the Pillars of Creation, and you know that at the quantum level particles are nonbinary—like many people. And you've heard of Onesimus, the Black African victim of kidnapping and slavery, who nonetheless made a significant contribution to early colonial medical knowledge by teaching his captors about smallpox inoculations. You know that there are Native Hawaiians thinking about "pono science" and fighting to save the land—and science—from colonialism.

I also hope you know, if you are from a community that people claim doesn't have an affinity for science, that you are in fact from a community that has an affinity for science. The people who see it otherwise are full of shit. And if you are a stay-at-home parent, I hope you know that your caring work is work, with social and economic value.

If you are someone who experiences anxiety at the thought of math or science, then you have been enormously brave in reading my book. I hope you have found it useful, and I am grateful to you for giving it some of your spacetime. Now that you know that you are an enormously brave person, I hope that you will protect that capacity and also put it to some good use. If we all work together, we can turn our freedom dreams into brilliant futures.

—Chanda Prescod-Weinstein, November 1, 2021

Acknowledgments

THOUGH MY NAME IS ON THE COVER, I KNOW THAT NO ONE THINKS alone. I have tried to write in rhythm with the ancestors. To the kiaʻi, past and present: thank you for the solidarity, the lessons, and for pointing us toward the future.

I am grateful that I've spent the first thirty-eight years of my life learning tenacity from Margaret Prescod, the most influential thinker in my life. On the first day of this journey, she made sure we had a midwife to defend us. Thank you, Mom, for keeping us alive and for showing me daily what it looks like to live your life knowing that it matters. Many other people helped raise me and shape the way I think, including my dad Sam, my stepmother Maria, my stepfather Maarten, my Grandma Elsa z"l, my Grandpa Norman z"l, my grandmother Selma, my Auntie Roz and Uncle George, and my Uncle Peter and Aunt Ina. I am thankful for Jasmine Young, who was my second Los Angeles mother. I am grateful to everyone in the Blackman-James-Prescod-Maldonado-McNabb-Paul-Preudhomme-Oostendorp-Weinstein-Williams-Ho-Tiu families for the ways you supported me as a baby scientist and an adult navigating the world. I am grateful for my sister Maya and the many cousins who are in my life, including Khari, Lisa, Sonya P., Sonya B., Ryan, Caleb, Jessica, Sean, Michael, Karen, Corina, Chanda, Barbara, and Lyndon. Thank you to Sharyll Burroughs and Devin Thomas, and Debbie, Jerry, Danny, Frances, and Marina Acosta for being part of the familial constellation.

Special thanks to Sharifah Zainab Williams, for not only being my best friend, my sister from another mister, and the person who is responsible for making sure I got the cover of this book right, but also for drawing nearly all the figures and bringing Gravity Girl into my life.

Thank you to the people in my life who are like family. Much gratitude in no particular order to Meredith Schweig, Andres Su, Sarah Tuttle, Alexis Shotwell, Sujay Ashok, Eleonora Dell'Aquila, Nikolaki (Νικολάκης) Konidaris, Vinny Manoharan, Sophia Cisneros, Robert Jones Jr., Marvin Pittman, Brian Shuve, Ayinde Jean-Baptiste, John Tsou, Daniel Wagner, Lawrence Ware, Elizabeth Crane, Kiese Laymon, Matthew Anderson, Corey Welch, Kyle Hernandez, A. J. Walker, John Jacobs, Stephani Page, Bakari Middleton, Tina Chong (we made it!), Helen Kim, Silvabug, David Sanders, Janice Chen, the Mighty Unicorns of the Los Angeles Center for Enriched Studies, Derek Larsen, Luke Winstrom, Alejo Stark, Michael Gould-Wartofsky, Cappy, Elizabeth Capuano, Shanequa Gay, and Richard Folkerts z"l.

Thank you to the many friends I've made on this journey and apologies to anyone I forget to name: The 'Nati, Astro AntiFa, P4J, the Trapp-Jordan family/collective, Toby Anekwe, Kamala Avila-Salmon, Moya Bailey, Grey Batie, Kabria Baumgartner, Brian Beckford, Ed Bertschinger, Mildred Boveda, Kinitra Brooks, Mahogany L. Browne, Matthew Buckley, Erica Buddington, Adam Bush, Ray Burks, Regina Caputo, Michelle Carter, Iokpea Casumbal-Salazar, Chang Jui-chuan, Linda Chavers, Ed Childs, Cam Moultrie Clemmons, Rabbi Shahar Colt, Kyle Cranmer, Djuna Croon, Chris Dixon, Keolu Fox, Aricka Foreman, Aria Halliday, Lawni Hamilton, Daniel Harlow, Narinda Heng, K. Renee Horton, Monica Huerta, Anton Hur, Seyda Ipek, Jedidah Isler, Lacy M. Johnson, Rhonda Shusli Baseler Johnson and Eugene Johnson, Vijay Iyer, Kay Kirkpatrick, Reba Kim, Rabbi Sandra Lawson, Mina Le, Danielle Lee, Franklin Leonard, Sean Lim, Elena Long, Peter Mabrucco, Uahikea Maile, Joseph Martin, K. Eric Martin, Jessica Massa, Sam McDermott, Ben McKean, Steven McPherson, Jami Valentine Miller, Matthew Morrison, Brian Nord, Okwudiri Onyedum, Charee Peters, Vaughn Rasberry, Matthew Reece, Joshua Roebke, Arianne Shahvisi, James Saliba, Marjorie Salvodon, Pearl Sandick, Katelin Schutz, Nausheen Shah, Kate Sim, Tracy Slatyer, Stephen Smith, SA

Smythe, Dean Spade, Rabbi Toba Spitzer, Jeremy Steffler, Tim Tait, Sadena Thevarajah, Johnny Trinh, Eve Tuck, Jessica Tucker, Natali Valdez, Lucianne Walkowicz, Elissa Washuta, Liz Wayne, Steve Weinstein, Graham White, Ustadza Williams, Caleph Wilson, Selam Woldeselassie, Xine Yao, Tien-Tien Yu, Mark Zastrow, and everyone at Congregation Dorshei Tzedek. Thanks to Emily Spieler and Gregory Wagner for the warm welcomes. Gratitude to Beth Salvia, Desirae Rodgers, Connor Trinneer, Gates McFadden, Rebekah Starks, Mahogany Harris, Barry Rice, Sonia Walker, Emily Tepper, Kendra James, and all the other Trekkies and Trek-makers who have shared in my enthusiasm for *Star Trek*. Particular thanks to Bryan Kamaoli Kuwada for the beautiful lessons.

Without my Los Angeles Unified School District teachers, especially Mr. Frank Wilson, Mr. Warren Buckner z"l, Mr. Randall Rutschman, Mr. Jed Laderman, and Ms. Elaine Berman z"l, I would not be the student of the universe that I am today. I am also grateful to the body workers who taught me and helped heal me, particularly Pam Garcia and Patrishya Fitzgerald, and the many *extraordinary* dental and medical professionals who helped me manage my disabilities over the years. Special thanks to my colleagues at the University of New Hampshire who gave me room to write this book, to the late Julie Williams z"l, and to all of my academic mentors and friends, especially Alan Guth, David Kaiser, Priya Natarajan, Risa Wechsler, Anna Watts, Karen Barad, Nadya Mason, Sekazi Mtingwa, Airea Matthews, Imani Perry, Christina Sharpe, Henry Frisch, Martin Elvis, Jonathan McDowell, Anthony Aguirre, Melissa Franklin, Richard Easther, Lee Smolin, Clifford Johnson, Julianne Dalcanton, Paul Gueye, Wesley Harris, and with a deep wish that they were around to read it, particle physicist Ann Nelson z"l, and Teamster and astronomer John Huchra z"l.

I am especially indebted to the people who read a draft or talked through ideas in it: Elizabeth Crane, Stephani Page, Bryan Kamaoli Kuwada, Arianne Shahvisi, Brian Shuve, Ashon Crawley, Aleks Diamond-Stanic, Autumn Kent, Sarah Tuttle, Robert McNees, Alexis Shotwell, Ryan Mandelbaum, Jose Osmundson, Eve Tuck, Ayinde Jean-Baptiste, and Kiese Laymon. Thank you to Rabbi Toba Spitzer for helping with the prayer selection and interpretation.

I am extremely grateful to my agent Jessica Papin and my editor Katy O'Donnell for believing that the real me was enough and the book I wanted to write was enough. Thanks to the wonderful team at Bold Type and Hachette, Lindsay Fradkoff, Jocelynn Pedro, Claire Zuo, Miguel Cervantes, the heroic Brynn Warriner and the production and printing press staff, for all your hard work. I also thank the many people who kept my living and workspaces clean, organized, and functional over the last twenty years, including those at the Aspen Center for Physics and the Kavli Institute for Theoretical Physics. I have depended on janitors, housekeeping, garbage collectors, dining service workers, food delivery workers, farmworkers, and administrative staff.

I thank Bitch Media for permission to use parts of my original Winter 2017 publication of "The Physics of Melanin," to the stunningly talented Lena Blackmon for permission to use her poem which first appeared in *The Offing*, and to Tim M. P. Tait for the updated Venn diagram. Thank you to Erica Buddington for her thoughts on night sky access.

Thank you to Alexis Ellison for soul-nourishing partnership, love, and for all the things we know but cannot say. We are the ones we've been waiting for.

To my spouse and political partner who is publicly known as MrProfChanda: Thank you for the phrase "All Our Galactic Relations." None of this would have been possible without your willingness to teach me about the law and ethnic studies, and your steadfast commitment to justice and integrity, sometimes at great personal expense. You believed in me when I didn't and coached me when I needed coaching. You listened, talked through my ideas with me, carefully edited chapters, took notes while I thought out loud in the shower, fed me, made sure I drank water, and took care of housework so that I could write in as little physical pain as possible. Your housework made this book possible. Also, just so you know, people can make friends on OkCupid.

A Reader's Guide to *The Disordered Cosmos*

By Shirley Ngozi Nwangwa

A Conversation with Chanda Prescod-Weinstein

You write early on in the book that "the universe is wonderfully queerer than we think." Who were you envisioning as an audience for this?

I was writing to my seventeen-year-old self. I realized I was queer right before my eighteenth birthday. Before I went off to college I really didn't know, even though I was a founding member of my high school's Gay-Straight Alliance. I thought I was just being a really good ally. LOL.

I knew there would be another queer Black girl who needed a book that was honest about how shit everything is. But that said: this is also *yours*. That is what I've wanted this book to do. There is nothing I can do about what happened to me, the things that I've gone through. But hopefully the knowledge and understanding I've gained by living through the things I've lived through mean that I can give someone else better tools to navigate similar experiences. Maybe I can even help radicalize them by saying, *Think bigger than just about making space in the room, think about whether this room should be here and what shape it should have and who should be in charge of it.* I wanted to create for the person who might be like me in some ways and unlike me in other ways. I hoped a lot of people would read the book, that it'd be valuable to different people for different reasons. But I think at its core that the book is for Black girls and femmes who like science.

I'd be lying if I didn't tell you that the chapter title "Spacetime Isn't Straight" really sent me. I yelled out, YES! I already knew that this book was going to be gay as hell. It's also really funny.

Physicists like wordplay. You see a lot of that in the book. That's cultural, it's part of physicist culture. It just happens to be that because I am different from other physicists, my wordplay is a little bit different, too. I thought of "spacetime isn't straight" as a joke and then texted Brian Shuve, a white, gay theoretical physicist who is a member of Particles for Justice with me. I said, "I can't really name a chapter this way, can I?" And he said, "You 100 percent can." This is what happens when queer physicists talk to each other: we give each other permission to do things we might not feel we had permission to do otherwise. And even I had to have that experience as part of the making of this book.

Brian was a really important reader for the book. Very important. All of my readers were. One of the reasons is that he was trained more traditionally than I was, as a particle physicist at Harvard. He knew all of the players. He knew there were things I'd said, that he'd heard me say, that I'd left out of the early draft he read, and he'd suggest that I add those things in. The kind of reader you want when you're writing a book is someone who will tell you about yourself—in a good way, too! I have a whole team around me, giving me permission.

Where does the "Dreams Deferred" part of the book's subtitle come from?

The Disordered Cosmos was the name of my blog in graduate school, and I got that name from a characterization of my first peer-reviewed scientific paper. You can have these non-local connections between different points in spacetime, so the connections wouldn't be orderly; they would be disordered.

I think a lot in terms of Langston Hughes, lines from Langston Hughes, which shows in the chapter titles. "Dreams Deferred," in reference to Hughes's *Montage of a Dream Deferred*, was important to include in the subtitle because it is recursive. The book represents a dream deferred, the book is about dreams deferred, the book is about challenging structures that defer dreams. The book was aimed at the end of dreams deferred as a phenomenon for Black people.

I am a literary reader at the end of the day. Look, I found a way to mention *Mansfield Park* in a book about physics. I wanted the title of the book to sound literary; it mattered to me that the title had craft in it. I wanted to do more than indicate what the book is about. I wanted it to have poetry in it.

I found it devastating to read in the book, "There have been times when I felt like the astro/physics community was actively ruining the stars for me because of its unwillingness to detach from dehumanizing ways of making science happen." I couldn't fathom what it might be like to feel as if your colleagues, all those guarding the gates, had the power to extinguish something as magnanimous as the stars.

It happens for a lot of people. We see it in the stats, we see people coming in, being excited to be Black and STEM, and then by the end of frosh year they're not majoring in STEM majors anymore. We see it with the people who say, "I'm going to be a physicist," and then graduate with their degree and don't go to graduate school, not because they found something else they love, but because they found something that doesn't hurt as much. This is the real problem: Black women are walking away because it hurts, not because they fell in love with something else. It's okay to fall in love with something else, and it's okay to have multiple lives. What's not okay is for the thing you love to hurt you. And the problem is that physics hurts and abuses people. I've written a whole chapter about how rape is part of the scientific story.

You've said that you went back and forth about writing that chapter, and that you were nervous that it would speak louder than everything else, by virtue of the subject matter.

I think that I've successfully scared reporters with that comment. That rhetorical trick worked! People have been really hesitant to ask me about that chapter, and really respectful. I have had to live through a lot of audacity regarding my rapist. In some sense, the chapter being there has become a solace; I ended up needing the chapter to be there. Because at least I told my story. But it has been a hardening experience. For the first few months, every time the chapter would come up, I would become dysfunctional for the rest of the day and there would be a lot of crying. Now I can just talk about it and it's like, yeah, that happened, and I'm happy and at least he's aging horribly, he looks terrible [*laughs*]. Survivors—I think that we're entitled to be petty. We're not entitled to take it out on other people, but we're entitled to be petty, especially if the subject of our pettiness is the perpetrator. I have heard from people that the chapter was useful to them, and I think that Lacy M. Johnson, author of *The Reckonings: Essays on Justice for the Twenty-First Century*, was right when she told me that writing it would give me control over my own story. I am glad I listened to her. I don't feel shame or sadness. It comes back to having people around you.

How much do you think comprehending science matters to your readers? How were you thinking about this as you wrote?

The universe is queer. It's really, really weird. If this stuff doesn't feel intuitive, that's okay. It's not intuitive for any of us—except maybe for drag performer and writer Amrou Al-Kadhi, who said, "Particles are themselves nonbinary," like nonbinary people. So, that shit is intuitive for Amrou, at least.

This is something I address in my note to the reader, an addition to the paperback: I think the thing that I did not think about enough—and I will in future, this has been a learning experience for me—was how much trauma and anxiety people were going to bring to the science discussion. I am so grateful that people who have been told that they were bad at math, who had horrific experiences in their high-school physics classes, or who were told that they were not smart enough to take the physics class, picked up the book anyway. I have a deep appreciation for those readers in particular because I know that picking up this book was not a simple thing. I did not think about the fact that people are socialized to think that a book about science will have a test at the end. Maybe the test isn't in the book, but the test is out there, lurking somewhere, and you're either going to pass the test or fail the test. I don't think I did a good job in the introduction of setting readers up for a different experience. This is a genuine error that I made: I didn't take into account that I had to tell people there was no test, and that it was okay for them not to memorize what I was saying about science.

As I write in my note to the reader, I bet anyone who reads this book has some sense of what a quark is now, but can they write out the Lagrangian for the Standard Model of particle physics? No? Well, I also can't do that off the top of my head. I mostly know one term from it, and it's the term that's related to my work. I have twenty years of training under me. If you expect that you're going to be as good at this after 150 pages as I am after twenty years, either your expectation of yourself is too high or you think I'm not very good at my job [*laughs*].

People will tell me that the science stuff went over their head, that they don't think they really got it, but then make jokes with me about quarks and fundamental particles. And I'm like, *So you know about the standard model, you know that spacetime isn't straight, you know that Black people are luminous matter, you know there are quarks, you know there are neutrinos, you know there's this weird concept of spin, you know these things are out there! Are you the world's greatest expert? No? But who the fuck cares, that wasn't the point.*

How do you want this book to fit into academic study?

I hope that it can be used as an example of how we can rethink science. The book is meant to be a holistic look at doing physics from the point of view of a Black, queer particle cosmologist. I hope it encourages people to rethink the established norms around all of those words, in some sense. I'm not primarily writing for academics. My intended audience, beyond my seventeen-year-old self, is anyone who's been told physics wasn't for them. This is an anti-gaslighting book. Physics is for you.

Questions for Thought and Discussion

1. What are your thoughts on the author's suggestion that the term "dark matter" be replaced by "invisible matter"? Do other stand-ins come to mind? Should the term be left as is?
2. Consider the opening to the chapter "Spacetime Isn't Straight" (pg. 45):

   > *As a physics student I was told repeatedly that I intuitively experienced space and time as completely different phenomena. As a physicist I wonder if that was ever true or just physics professors projecting their experiences onto us undergrads. In reality, I don't think I ever gave it much thought before I was told what to think about it.*

   What kind of harm is Prescod-Weinstein suggesting might occur when professors project their own experiences onto students? What effect might being taught the "right" way to think have on a young person?

3. What makes the concept of weak gravitational lensing an effective analogy for systemic racism? Where does the analogy falter? Are these kinds of analogies necessary?
4. One definition of freedom to which Prescod-Weinstein subscribes comes from the artist Shanequa Gay, who told the author, "Freedom looks like choice making without having to consider so many others when I make those choices" (pg. 7). What is your definition of freedom, and does Gay's sentiment resonate with yours?
5. What is another subject, outside of science, that you would like to see "queered" and why?
6. Prescod-Weinstein explains that gravity "is not a real force" (pg. 56), in the sense that what is actually occurring is the curving of spacetime. How does this sit with you? Does this point of view have the capacity to change the way you interact with the world or do you feel entirely unimpacted by it?
7. Throughout *The Disordered Cosmos*, Prescod-Weinstein uses her own Blackness, queerness, and relationship to womanhood as vehicles in presenting a holistic perspective on physics and cosmology. Discuss the idea that the human, the personal, cannot be taken out of science. What do you think about the notion that even science writing that positions itself as detached necessarily incorporates the writer's identity?
8. The author describes the feeling of having her love of the stars actively ruined by those unwilling to "detach from dehumanizing ways of making science happen" (pg. 156). If you're comfortable with sharing: Is this an experience that's familiar to you? If so, how might the experience have impacted your work, your health, your joy?
9. Apart from the physics of melanin and weak gravitational lensing, which scientific topics—in this book or outside of it—can you imagine as being covered in the hypothetical science section of a magazine geared toward Black readers?
10. What makes physics, and science more generally, political? What makes it colonialist?
11. Did *The Disordered Cosmos* make you more trusting or more skeptical of science, the history of science, and the capacity of scientists to stick to their own standards? Why?

Chanda Prescod-Weinstein's *Disordered Cosmos* Playlist

"Lift Every Voice and Sing" (Live), Kim Weston
"Final Form," Sampa the Great
"Boot," Tamar-Kali
"Balas y Chocolate," Lila Downs
"Spinning Wheel" (Remastered), Dr. Lonnie Smith
"Many Moons," Janelle Monáe
"Cold War," Janelle Monáe
"What Now," Rihanna
"I Wish I Knew How It Would Feel to Be Free," Nina Simone
"Redemption Song," Bob Marley & The Wailers
"I Didn't Know My Own Strength," Whitney Houston
"How I Got Over," The Roots
"A Love Supreme, Pt. IV—Psalm," John Coltrane
"Sally Ride," Janelle Monáe
"Tell Me No," Whitney Houston
"GhettoMusick," Outkast
"Overcome" (feat. Nile Rodgers), Laura Mvula, Nile Rodgers
"Dirty Gold," Angel Haze
"We Here Now," MC Lyte, Mario
"What I Did for Love," Emeli Sandé, David Guetta
"Lift Every Voice and Sing," Art Blakey & The Jazz Messengers

Recommended Reading

Books are my love language. In the preceding chapters, I have mentioned many books and texts that contribute to the development of my ideas. I have tried to list the texts I relied on most heavily below. I have certainly forgotten influential works, and I apologize to the authors. There is of course much more to read beyond what I have used. Many of these books and articles can be found through your local library. If they aren't available, let me just say that librarians and archivists are often the very, very best of humanity. This book wouldn't exist without them, and I bet they'd be happy to help you get your hands on the book or article you're looking for.

In the Beginning: A Bedtime Story

Gross, Alan G. *The Scientific Sublime: Popular Science Unravels the Mysteries of the Universe*. Oxford: Oxford University Press, 2018.

Sagan, Carl. *Cosmos*. New York: Random House, 1980.

Chapter One: I ♥ Quarks

Barad, Karen. *Meeting the Universe Halfway: Quantum Physics and the Entanglement of Matter and Meaning*. Durham: Duke University Press, 2007.

Barad, Karen. "No Small Matter: Mushroom Clouds, Ecologies of Nothingness, and Strange Topologies of SpaceTimeMattering." *Arts of*

Living on a Damaged Planet: Ghosts and Monsters of the Anthropocene, edited by Anna Tsing, Heather Swanson, Elaine Gan, and Nils Bubandt, 103–120. Minneapolis: University of Minnesota Press, 2017.

Hawking, Stephen. *A Brief History of Time: From the Big Bang to Black Holes*. New York: Bantam Dell, 1988.

Johnson, Lacy M. *The Reckonings*. New York: Scribner, 2018.

Joyce, James. *Finnegans Wake*. London: Faber and Faber, 1939.

Riordan, Michael. *The Hunting of the Quark: A True Story of Modern Physics*. New York: Simon & Schuster, 1987.

Other Suggested Reading

Aguirre, Anthony. *Cosmological Koans: A Journey to the Heart of Physical Reality*. New York: W.W. Norton Company, 2019.

Carroll, Sean. *The Particle at the End of the Universe: How the Hunt for the Higgs Boson Leads Us to the Edge of a New World*. New York: Dutton, 2012.

Johnson, Clifford V. *The Dialogues: Conversations about the Nature of the Universe*. Cambridge: MIT Press, 2017.

Mack, Katie. *The End of Everything (Astrophysically Speaking)*. New York: Scribner, 2020.

Smolin, Lee. *Einstein's Unfinished Revolution: The Search for What Lies Beyond the Quantum*. New York: Penguin Press, 2019.

Chapter Two: Dark Matter Isn't Dark

Natarajan, Priyamvada. *Mapping the Heavens: The Radical Scientific Ideas That Reveal the Cosmos*. New Haven: Yale University Press, 2016.

Rubin, Vera C. *Bright Galaxies, Dark Matters*. New York: Copernicus Books, 1996.

Chapter Three: Spacetime Isn't Straight

Craig, John. "Isaac Newton and the Counterfeiters." *The Royal Society Journal of the History of Science* 18, no. 2 (December 1963): 136–145.

Edwards, Sue Bradford, and Duchess Harris. *Hidden Human Computers: The Black Women of NASA*. Minneapolis: Essential Library, 2016.

Einstein, Albert. *The Principle of Relativity*. North Yorkshire: Methuen and Company, 1923.

Green, Lesley, and David R. Green. *Knowing the Day, Knowing the World: Engaging Amerindian Thought in Public Archaeology*. Tucson: University of Arizona Press, 2013.

Sagan, Carl. *Billions and Billions: Thoughts on Life and Death at the Brink of the Millennium*. New York: Random House, 1997.

Shetterly, Margot Lee. *Hidden Figures: The American Dream and the Untold Story of the Black Women Who Helped Win the Space Race*. New York: William Morrow and Company, 2016.

Smolin, Lee. *Three Roads to Quantum Gravity*. New York: Basic Books, 2001.

Wald, Robert. *General Relativity*. Chicago: University of Chicago Press, 1984.

Chapter Four: The Biggest Picture There Is

Guth, Alan. *The Inflationary Universe: The Quest for a New Theory of Cosmic Origins*. New York: Perseus Books, 1997.

hoʻomanawanui, kuʻualoha. *Voices of Fire: Reweaving the Literary Lei of Pele and Hiʻiaka*. Minneapolis: University of Minnesota Press, 2014.

Lineweaver, Charles. "Inflation and the Cosmic Microwave Background." *The New Cosmology: Proceedings of the 16th International Physics Summer School*, edited by M. M. Colless. Singapore: World Scientific, 2005.

Mickens, Ronald E. "The Life and Work of Elmer Samuel Imes." *Physics Today* 70, no. 10 (October 2018): 28–35.

Steele, Julia. "The Meaning of Aloha ʻĀina with Professor Jon Osorio." *Aloha Aina*. Hawaiʻi Public Radio. February 5, 2016.

Chapter Five: The Physics of Melanin

Benjamin, Ruha. *Race After Technology: Abolitionist Tools for the New Jim Code*. Cambridge: Polity, 2019.

Giacomantonio, Clare. *Charge Transport in Melanin, a Disordered Bio-Organic Conductor*. Undergraduate Thesis, University of Queensland, 2005.

Hobson, Janell. *Venus in the Dark: Blackness and Beauty in Popular Culture*. London: Routledge, 2018.

Jordan, June. "Poem for South African Women." *Passion: New Poems, 1977–1980*. Boston: Beacon Press, 1980.

McLean, Shay-Akil. "Social Constructions, Historical Grounds." *Practicing Anthropology* 42, no. 3 (Summer 2020): 40.

Millington, George W.M., and Nick J. Levell. "From Genesis to Gene Sequencing: Historical Progress in the Understanding of Skin Color." *International Journal of Dermatology* 46, no. 1 (January 3, 2007): 103–105.

Roberts, Dorothy. *Fatal Invention: How Science, Politics, and Big Business Re-create Race in the Twenty-First Century*. New York: New Press, 2011.

Westerhof, Wiete. "The Discovery of the Human Melanocyte." *Pigment Cell & Melanoma Research* 19, no. 3 (May 16, 2006): 183–193.

Womack, Ytasha L. *Afrofuturism: The World of Black Sci-Fi and Fantasy Culture*. Chicago: Lawrence Hill Books, 2013.

Chapter Six: Black People Are Luminous Matter

Chavers, Linda. "Here's My Problem With #BlackGirlMagic: Black Girls Aren't Magical. We're Human." *Elle*. January 13, 2016.

Du Bois, W. E. B. *The Souls of Black Folk: Essays and Sketches*. Chicago: A. C. McClurg & Co., 1903.

Ellison, Ralph. *Invisible Man*. New York: Random House, 1952.

Thomas, Sheree R. *Dark Matter: A Century of Speculative Fiction from the African Diaspora*. New York: Aspect, 2000.

Chapter Seven: Who Is a Scientist?

Austen, Jane. *Mansfield Park*. London: Thomas Egerton, 1814.

Baptist, Edward. *The Half Has Never Been Told: Slavery and the Making of American Capitalism*. New York: Basic Books, 2013.

Brock, Claire. *The Comet Sweeper: Caroline Herschel's Astronomical Ambition*. London: Icon Books, 2004.

Douglass, Frederick. *Narrative of the Life of Frederick Douglass, an American Slave*. Boston: Anti-Slavery Office, 1845.

Haley, Alex. *Roots: The Saga of an American Family*. New York: Doubleday, 1976.

Perry, Imani. *Vexy Thing: On Gender and Liberation*. Durham: Duke University Press, 2018.

Robinson, Cedric J. *Black Marxism: The Making of the Black Radical Tradition*. Chapel Hill: University of North Carolina Press, 1983.

Simmons, LaKisha. *Crescent City Girls: The Lives of Young Black Women in Segregated New Orleans*. Chapel Hill: University of North Carolina Press, 2015.

Truth, Sojourner. *Narrative of Sojourner Truth: A Northern Slave*. The Author, 1850.

Walker, Alice. "Saving the Life That Is Your Own." *In Search of Our Mothers' Gardens*. San Diego: Harcourt Brace Jovanovich, 1983.

Winterburn, Emily. *The Quiet Revolution of Caroline Herschel: The Lost Heroine of Astronomy*. Brimscombe: History Press, 2017.

Chapter Eight: Let Astro/Physics Be the Dream It Used to Be

Aitkenhead, Decca. "Peter Higgs: I Wouldn't Be Productive Enough for Today's Academic System." *The Guardian*. December 6, 2013. www.theguardian.com/science/2013/dec/06/peter-higgs-boson-academic-system.

Kahn, Jonathan. *Race on the Brain: What Implicit Bias Gets Wrong About the Struggle for Racial Justice*. New York: Columbia University Press, 2017.

Lorde, Audre. "The Uses of Anger: Women Responding to Racism." *Women's Studies Quarterly* 9, no. 3 (1981): 7–10.

Smolin, Lee. *The Trouble With Physics: The Rise of String Theory, the Fall of a Science, and What Comes Next*. Boston: Houghton Mifflin Harcourt, 2006.

Wynter, Sylvia, and McKittrick, Katherine. "Unparalleled Catastrophe for our Species? Or, to Give Humanities a Different Future: Conversations." In *Sylvia Wynter: On Being Human as Praxis*, edited by Katherine McKittrick. Durham: Duke University Press, 2015, 9-89.

Chapter Nine: The Anti-Patriarchy Agender

American Physical Society. *LGBT Climate in Physics: Building an Inclusive Community*. 2016. www.aps.org/programs/lgbt/upload/LGBTClimateinPhysicsReport.pdf.

Bailey, Moya. *Misogynoir Transformed: Black Women's Digital Resistance*. New York: NYU Press, 2021.

Bailey, Moya. "More On the Origin of Misogynoir." *Moyazb*. April 27, 2014. https://moyazb.tumblr.com/post/84048113369/more-on-the-origin-of-misogynoir.

Bailey, Moya. "They Aren't Talking About Me . . ." *The Crunk Feminist Collective.* March 14, 2010. www.crunkfeministcollective.com/2010/03/14/they-arent-talking-about-me.

Bey, Marquis. *Anarcho-Blackness: Notes Toward a Black Anarchism.* Chico: AK Press, 2020.

Cava, Peter. "Activism, Politics, and Organizing." *Trans Bodies, Trans Selves: A Resource for the Transgender Community*, edited by Laura Erickson-Schroth. Oxford: Oxford University Press, 2014.

De La Rosa, Ralph. *Don't Tell Me To Relax: Emotional Resilience in the Age of Rage, Feels, and Freak-Outs.* Boulder: Shambhala Publications, 2020.

Haritaworn, Jin, and C. Riley Norton. "Trans Necropolitics: A Transnational Reflection on Violence, Death, and the Trans of Color Afterlife." In *The Transgender Studies Reader 2*, edited by Susan Stryker and Aren Z. Aizura. New York: Routledge, 2013, 66–76.

Harris-Perry, Melissa. *Sister Citizen: Shame, Stereotypes, and Black Women in America.* New Haven: Yale University Press, 2011.

Jackson, Zakiyyah Iman. *Becoming Human: Matter and Meaning in an Antiblack World.* New York: NYU Press, 2020.

Muñoz, José Esteban. *Cruising Utopia: The Then and There of Queer Futurity.* New York: NYU Press, 2009.

Prescod-Weinstein, Chanda. "Making Black Women Scientists Under White Empiricism: The Racialization of Epistemology in Physics." *Signs: Journal of Women in Culture and Society* 45, no. 2 (2020): 421–447.

Shotwell, Alexis. *Knowing Otherwise: Race, Gender, and Implicit Understanding.* University Park, PA: Penn State University Press, 2011.

Spade, Dean. "The Right Wing Is Leveraging Trans Issues to Promote Militarism." *Truthout.* April 5, 2017. https://truthout.org/articles/the-right-wing-is-leveraging-trans-issues-to-promote-militarism.

Trudy. *Gradient Lair.* October 22, 2013. www.gradientlair.com/post/64818257366/farryn-johnson-blonde-hair-hooters-fired.

Chapter Ten: Wages for Scientific Housework

James, Selma. *Sex, Race, and Class—The Perspective of Winning: A Selection of Writings 1952–2011.* Oakland: PM Press, 2012.

New York Wages for Housework Committee. *Wages for Housework.* Mid-1970s. http://bcrw.barnard.edu/archive/sexualhealth/WagesForHousework.pdf.

Prescod-Weinstein, Chanda. "Surviving and Thriving as an Underrepresented Minority Astro/Physics Student, part 1." *Medium.*

September 9, 2015. https://medium.com/@chanda/surviving-and-thriving-how-to-be-a-urm-astro-physics-student-part-1-97df8e81eb59.

Traweek, Sharon. *Beamtimes and Lifetimes: The World of High Energy Physicists*. Cambridge: Harvard University Press, 1988.

Chapter Eleven: Rape Is Part of This Scientific Story

Driscoll, Frances. *The Rape Poems*. Seattle: Pleasure Boat Studios, 1997.

Potter, Elizabeth. *Gender and Boyle's Law of Gases*. Bloomington: Indiana University Press, 2001.

Prescod-Weinstein, Chanda. "Diversity Is a Dangerous Set-up." *Medium*. January 24, 2018. https://medium.com/space-anthropology/diversity-is-a-dangerous-set-up-8cee942e7f22.

Chapter Twelve: The Point of Science: Lessons from the Mauna

Green, Lesley J. F. "Challenging Epistemologies: Exploring Knowledge Practices in Palikur Astronomy." *Futures* 41, no. 1 (2009): 41–52.

Green, Lesley J. F. "'Indigenous Knowledge' and 'Science': Reframing the Debate on Knowledge Diversity." *Archaeologies* 4 (2008): 144–163.

Hopkinson, Nalo, and Uppinder Mehan, eds. *So Long Been Dreaming: Postcolonial Science Fiction and Fantasy*. Vancouver: Arsenal Pulp Press, 2004.

Jeffries, Stuart. "Anarchy at the South Pole: Santiago Sierra Plants the Black Flag to Destroy All Borders." *The Guardian*. September 25, 2018. www.theguardian.com/artanddesign/2018/sep/25/santiago-sierra-south-pole-anarchy-flags-spanish-artist-dundee-synagogue.

LaDuke, Winona. *All Our Relations: Native Struggles for Land and Life*. Boston: South End Press, 2008.

Lyons, Scott Richard. "Rhetorical Sovereignty: What Do American Indians Want from Writing?" *College Composition and Communication* 51, no. 3 (2000): 447–468.

Sharpe, Christina. *In the Wake: On Blackness and Being*. Durham: Duke University Press, 2016.

Smith, A. Mark. *From Sight to Light: The Passage from Ancient to Modern Optics*. Chicago: University of Chicago Press, 2014.

Tuck, Eve, and K. Wayne Yang. "Decolonization Is Not a Metaphor." *Decolonization: Indigeneity, Education & Society* 1, no. 1 (2012): 1–40.

Whitt, Laurelyn. *Science, Colonialism, and Indigenous Peoples: The Cultural Politics of Law and Knowledge.* Cambridge: Cambridge University Press, 2009.

Other Suggested Reading

Aikau, Hokulani K., and Vernadette Vicuña Gonzalez, editors. *Detours: A Decolonial Guide to Hawaiʻi.* Durham: Duke University Press, 2019.

Arvin, Maile. *Possessing Polynesians: The Science of Settler Colonial Whiteness in Hawaiʻi and Oceania.* Durham: Duke University Press, 2019.

Coffman, Tom. *Nation Within: The History of the American Occupation of Hawaiʻi.* Durham: Duke University Press, 2016.

Kauanui, J. Kēhaulani. *Paradoxes of Hawaiian Sovereignty: Land, Sex, and the Colonial Politics of State Nationalism.* Durham: Duke University Press, 2018.

Prescod-Weinstein, Chanda. "Decolonizing Science Reading List," Medium, April 25, 2015. https://medium.com/@chanda/decolonising-science-reading-list-339fb773d51f.

Chapter Thirteen: Cosmological Dreams Under Totalitarianism

Du Bois, W. E. B. *Black Reconstruction in America 1860–1880: An Essay Toward a History of the Part Which Black Folk Played in the Attempt to Reconstruct Democracy in America.* New York: Harcourt, Brace and Co., 1935.

Greenberg, Daniel S. *The Politics of Pure Science.* New York: New American Library, 1967.

Hughes, Langston. "Let America Be America Again." *The Collected Poems of Langston Hughes.* New York: Alfred A. Knopf, 1994.

Ignatin, Noel, and Ted Allen. *White Blindspot and Can White Radicals Be Radicalized.* Detroit: The Radical Education Project and New York: NYC Revolutionary Youth Movement, 1969.

James, C. L. R. *Mariners, Renegades and Castaways: The Story of Herman Melville and the World We Live In.* Detroit: Bewick Editions, 1953.

McClellan, James E, III. *Colonialism and Science: Saint Domingue and the Old Regime.* Chicago: University of Chicago Press, 2010.

McClellan, James E., III, and François Regourd. "The Colonial Machine: French Science and Colonization in the Ancient Regime." *Osiris* 15 (2000): 31–50.

Monk, Ray. *Robert Oppenheimer: His Life and Mind*. New York: Random House, 2013.

Prescod-Weinstein, Chanda, Sarah Tuttle, and Joseph Osmundson. "We are the Scientists Against a Fascist Government." *The Establishment*. February 2, 2017. https://medium.com/the-establishment/we-are-the-scientists-against-a-fascist-government-d44043da274e.

Rasberry, Vaughn. *Race and the Totalitarian Century: Geopolitics in the Black Literary Imagination*. Cambridge: Harvard University Press, 2016.

Wolfe, Audra. *Freedom's Laboratory: The Cold War Struggle for the Soul of Science*. Baltimore: Johns Hopkins University Press, 2019.

Chapter Fourteen: Black Feminist Physics at the End of the World

Big Door Brigade. "What Do We Mean by 'Mutual Aid'?" 2017. https://bigdoorbrigade.wordpress.com/2017/01/31/what-do-we-mean-by-mutual-aid.

Collins, Patricia Hill. *Black Feminist Thought: Knowledge, Consciousness, and the Politics of Empowerment*. London: Unwin Hyman, 1990.

The Combahee River Collective. "The Combahee River Collective Statement." *All the Women Are White, All the Blacks Are Men, But Some of Us Are Brave: Black Women's Studies*, edited by Akasha (Gloria T.) Hull, Patricia Bell-Scott, and Barbara Smith. New York: Feminist Press, 1982.

Generation Five. *Toward Transformative Justice: A Liberatory Approach to Child Sexual Abuse and Other Forms of Intimate and Community Violence*. 2007. www.usprisonculture.com/blog/wp-content/uploads/2012/03/G5_Toward_Transformative_Justice.pdf.

Goodyear-Ka'ōpua, Noelani. "Introduction." *A Nation Rising: Hawaiian Movements for Life, Land, and Sovereignty*. Edited by Noelani Goodyear-Ka'ōpua, Ikaika Hussey, and Erin Kahunawaika'ala Wright. Durham: Duke University Press, 2014.

Hartman, Saidiya. *Scenes of Subjection: Terror, Slavery, and Self-Making in Nineteenth-Century America*. Oxford: Oxford University Press, 1997.

Hayes, Chris. "Thinking About How to Abolish Prisons With Mariame Kaba." *MSNBC*. April 10, 2019. www.nbcnews.com/think/opinion/thinking-about-how-abolish-prisons-mariame-kaba-podcast-transcript-ncna992721.

Herzig, Rebecca. *Suffering for Science: Reasoning and Sacrifice in Modern America*. Rutgers: Rutgers University Press, 2005.

Johnson, Fenton. "The Future of Queer." *Harper's Magazine*. January 2018, 27–34.

Kaba, Mariame. "Yes, We Literally Mean Abolish the Police." *New York Times*. June 12, 2020. www.nytimes.com/2020/06/12/opinion/sunday/floyd-abolish-defund-police.html.

Lorde, Audre. *Sister Outsider: Essays and Speeches*. Berkeley: Crossing Press, 1984.

Maile, David Uahikeaikalei'ohu. "Science, Time, and Mauna a Wākea: The Thirty-Meter Telescope's Capitalist-Colonialist Violence, Part I." *Red Nation*. May 13, 2015. http://therednation.org/2015/05/13/science-time-and-mauna-a-wakea-the-thirty-meter-telescopes-capitalist-colonialist-violence-an-essay-in-two-parts.

Maile, David Uahikeaikalei'ohu. "Science, Time, and Mauna a Wākea: The Thirty-Meter Telescope's Capitalist-Colonialist Violence, Part II." *Red Nation*. May 20, 2015. https://therednation.org/2015/05/20/science-time-and-mauna-a-wakea-the-thirty-meter-telescopes-capitalist-colonialist-violence.

Martin, Joseph. *Solid State Insurrection: How the Science of Substance Made American Physics Matter*. Pittsburgh: University of Pittsburgh Press, 2018.

Nhất Hạnh, Thích. *The Heart of the Buddha's Teaching: Transforming Suffering into Peace, Joy, and Liberation*. Parallax Press, 1998.

Publia, Alexandre. "South African Students' Question: Remake the University or Restructure Society?" *Abolition Journal*. December 22, 2016. https://abolitionjournal.org/south-african-students-question-remake-the-university-or-restructure-society.

Rich, Adrienne. "Why I Refused the National Medal for the Arts." *Arts of the Possible: Essays and Conversations*. New York: W.W. Norton Company, 2001.

Shotwell, Alexis. *Against Purity: Living Ethically in Compromised Times*. Minneapolis: University of Minnesota Press, 2016.

Spade, Dean. "Trans Politics on a Neoliberal Landscape." Lecture. February 9, 2009. Barnard College. Lecture. http://bcrw.barnard.edu/videos/dean-spade-trans-politics-on-a-neoliberal-landscape.

Wilson, Shawn. *Research Is Ceremony: Indigenous Research Methods*. Nova Scotia: Fernwood Publishing, 2008.

Other Suggested Reading

Creative Interventions. "Creative Interventions Toolkit: A Practical Guide to Stop Interpersonal Violence." www.creative-interventions.org/tools/toolkit.

Dear Mama, This Is What My Freedom Dream Looks Like

Hendrix, Kathleen. "Passionate Pursuer's Crusade Against the South Side Slayer: Margaret Prescod Trying to Raise Community Awareness on the Streets of South-Central L.A . . . and Beverly Hills." *Los Angeles Times*. October 16, 1986. www.latimes.com/archives/la-xpm-1986-10-16-vw-5852-story.html.

Kelley, Robin D. G. *Freedom Dreams: The Black Radical Imagination*. Boston: Beacon Press, 2003.

Index

COURTESY OF THE AUTHOR

Chanda Prescod-Weinstein is an assistant professor of physics and astronomy and core faculty in women's and gender studies at the University of New Hampshire. She is also a columnist for *New Scientist* and *Physics World*. Her research in theoretical physics focuses on cosmology, neutron stars, and dark matter. She also does research in Black feminist science, technology, and society studies. *Essence* magazine recognized her as one of "15 Black Women Who Are Paving the Way in STEM and Breaking Barriers." She has been profiled in several venues, including *TechCrunch*, *Ms. Magazine*, *Huffington Post*, *Gizmodo*, *Nylon*, and the African American Intellectual History Society's *Black Perspectives*. A cofounder of Particles for Justice, she has received the 2017 LGBT+ Physicists Acknowledgement of Excellence Award for her contributions to improving conditions for marginalized people in physics and the 2021 American Physical Society Edward A. Bouchet Award for her contributions to particle cosmology. Originally from East L.A., she divides her time between the New Hampshire Seacoast and Cambridge, Massachusetts.